KEY THEOREMS AND DEFINITIONS

(continued)

[3.6] $(a + b)(c + d) = ac + ad + bc + bd$

       $(a + b)^2 = a^2 + 2ab + b^2$

       $(a - b)^2 = a^2 - 2ab + b^2$

       $(a + b)(a - b) = a^2 - b^2$

[3.7] $\dfrac{A + B + C + \cdots}{M} = \dfrac{A}{M} + \dfrac{B}{M} + \dfrac{C}{M} + \cdots \quad (M \neq 0)$

[4.5] $a^2 - b^2 = (a + b)(a - b)$

       $a^2 + 2ab + b^2 = (a + b)^2$

       $a^2 - 2ab + b^2 = (a - b)^2$

[4.6]   If $m \cdot n = 0$, then either $m = 0$, $n = 0$, or both.

[5.1] $\dfrac{m}{m} = 1 \quad (m \neq 0)$

       $\dfrac{a}{b} \cdot \dfrac{c}{d} = \dfrac{ac}{bd} \quad (b \neq 0, d \neq 0)$

       $\dfrac{a}{b} = \dfrac{a}{b} \cdot \dfrac{c}{c} = \dfrac{ac}{bc} \quad (b \neq 0, c \neq 0)$

       $\dfrac{a}{b} = \dfrac{a \div k}{b \div k} \quad (b \neq 0, k \neq 0)$

[5.3] $x \cdot \dfrac{1}{x} = 1 \quad (x \neq 0)$

       $\dfrac{a}{b} \div \dfrac{c}{d} = \dfrac{a}{b} \cdot \dfrac{d}{c} = \dfrac{ad}{bc} \quad (b \neq 0, c \neq 0, d \neq 0)$

[5.4] $\dfrac{a}{c} + \dfrac{b}{c} = \dfrac{a + b}{c} \quad (c \neq 0)$

[6.5]   If $A = B$ and $C = D$, then $A + C = B + D$.

[7.2] $\sqrt{ab} = \sqrt{a}\,\sqrt{b} \quad (a, b \geq 0)$

       $\sqrt{\dfrac{a}{b}} = \dfrac{\sqrt{a}}{\sqrt{b}} \quad (a, b > 0)$

[7.5] $(\sqrt{N})^2 = N \quad (N \geq 0)$

       If $x^2 = a$, then $x = \sqrt{a}$ or $x = -\sqrt{a} \quad (a \geq 0)$.

[7.6]   The solutions to $ax^2 + bx + c = 0 \quad (a \neq 0)$ are

$$x = \frac{-b \pm \sqrt{b^2 + 4ac}}{2a}$$

[8.1] $\text{slope} = m = \dfrac{\text{change in } y}{\text{change in } x} = \dfrac{y_2 - y_1}{x_2 - x_1}$

[8.2]   The equation of the line through (x_1, y_1) with slope m is given by

$$y - y_1 = m(x - x_1)$$

ELEMENTARY ALGEBRA

ELEMENTARY
ALGEBRA

Howard A. Silver

Department of Mathematics
Chicago State University

PRENTICE-HALL, INC., ENGLEWOOD CLIFFS, NEW JERSEY 07632

Library of Congress Cataloging in Publication Data

SILVER, HOWARD A., (date)
    Elementary algebra.

    Includes indexes.
    1. Algebra.   I. Title.
QA152.2.S563   512.9   81-15348
ISBN 0-13-252817-7   AACR2

Editorial/production supervision by Kathleen M. Lafferty
Manufacturing buyer: John B. Hall

 © 1982 by Prentice-Hall, Inc.
A Simon & Schuster Company
Englewood Cliffs, New Jersey 07632

Printed in the United States of America

10   9   8   7   6

ISBN 0-13-252817-7

Prentice-Hall International (UK) Limited, *London*
Prentice-Hall of Australia Pty. Limited, *Sydney*
Prentice-Hall Canada Inc., *Toronto*
Prentice-Hall Hispanoamericana, S.A., *Mexico*
Prentice-Hall of India Private Limited, *New Delhi*
Prentice-Hall of Japan, Inc., *Tokyo*
Simon & Schuster Asia Pte. Ltd., *Singapore*
Editora Prentice-Hall do Brasil, Ltda., *Rio de Janeiro*

To my family

CONTENTS

Chapter   **6**   **GRAPHING AND LINEAR SYSTEMS**   **199**

Chapter   **7**   **ROOTS AND RADICALS**   **245**

PREFACE

This text, *Elementary Algebra*, is written for college students who need algebra skills to proceed in their major fields of study. It is designed to be used in a one-semester course with a prerequisite of basic arithmetic skills. The approach of the text is skills oriented, since the students using it will be mainly business majors, health-science majors, life-science majors, education majors, social-science majors, and so on.

The text can be used in both a lecture or an individualized format. Some of the main features of the text include:

READABILITY   The text is written in simple, down-to-earth English. Concepts are explained clearly and are not discussed into the ground.

EXAMPLES   The key to learning mathematical skills is in the examples. Here, the examples are carefully chosen to illustrate the concepts. They are carefully explained with step-by-step flow charts beside the steps of the example.

PROBLEM SETS   Like the examples, the problems in each section are carefully chosen. In fact, the exercises in the problem sets are keyed to the examples. For instance, beside problems 10 to 18 might appear a box that reads, "See Example 12." This is an aid to direct the student back to the examples for help.

PROCEDURES   All of the procedures, rules, and formulas are set off on the page with a box for handy reference.

APPLICATIONS   At the end of almost every problem set are various applications of the skills to fields such as business, health, chemistry, life science, optics, psychology, physics, electricity, and so on. These applications serve to reinforce the material as well as to stress the relevance of the material.

HAND CALCULATOR   Whenever possible, use of the hand calculator has been employed. This can help to take the sting out of the harder calculations.

YES/No Boxes   These boxes appear throughout the entire text and serve to warn the student of potential danger in common mistakes. The mathematical statement is made in two forms: correctly, under the YES column; and incorrectly, under the NO column, with a big X through it.

REVIEW   Every chapter has a summary of important words, important properties, and important procedures. There is also a set of review exercises that prepare the student for an in-class test on the chapter's material.

ANSWERS   The answers to the odd-numbered problems in the problem sets as well as to all the answers from the review exercises appear at the back of the text.

I gratefully acknowledge the help of the following reviewers during the various stages in the preparation of the manuscript: Marna Belcher, University of Oregon; Ann Bretscher, University of Georgia; Faye Thames, Lamar University; Fred Toxopeus, Kalamazoo Valley Community College; Rosalyn Wells, Georgia Southern College; and Bennie Zinn, San Antonio College.

I would like to thank the great Prentice-Hall staff for all their help and encouragement in the making of this textbook: my production editor, Kathleen Lafferty; my acquisitions editor, Bob Sickles; my field editor, Cindy Wallech; and the other production and marketing people whose names I never learned. I would also like to thank Jackie Blackmon and Jeni Morison for reading and reviewing the manuscript and Cecelia Roberts for typing the manuscript. Finally, I want to thank the faculty, tutors, and students of Chicago State University who gave me feedback on this textbook while it was used in its penultimate form.

I hope that this text proves successful and helpful to the students who use it. I would greatly appreciate any feedback, written or oral, by students or by faculty, on this text. This will help me improve the text for later versions.

H. SILVER

1
BASIC CONCEPTS

SYMBOLS AND OPERATIONS

The major difference between algebra and arithmetic is algebra's use of letters to stand for unknown numbers. For instance, the expression $x + 4$ means "some number plus four." Until we know what x is, we must leave this as $x + 4$.

Let us discuss the standard symbols of algebra. First is the **equal sign.**

ALGEBRAIC SYMBOLS	ENGLISH MEANING
$x = y$	x is the same number as y
$x \neq y$	x is *not* the same number as y

EXAMPLE 1   The following are examples of $=$ and $\neq$.

ALGEBRA	ENGLISH
(a) $10 = 10$	10 is the same number as 10
(b) $2 + 3 = 5$	$2 + 3$ is the same number as 5
(c) $4 \neq 3$	4 is *not* the same number as 3
(d) $1 + 5 \neq 9$	$1 + 5$ is *not* the same number as 9

The statement that two numbers are equal is called an **equation.**

The other symbols we use in comparing numbers are the **inequality symbols.**

ALGEBRAIC SYMBOLS	ENGLISH MEANING
$x < y$	x is less than y
$x > y$	x is greater than y
$x \leq y$	x is less than or equal to y
$x \geq y$	x is greater than or equal to y

To help remember the meaning of these symbols, notice that the smaller pointed ends of $<$ and $>$ always point to the smaller number, while the larger open ends point to the larger number.

EXAMPLE 2   The following examples use the inequality symbols.

ALGEBRA	ENGLISH
(a) $2 < 5$	2 is less than 5
(b) $15 > 11$	15 is greater than 11
(c) $8 \leq 10$	8 is less than or equal to 10
(d) $7 \leq 7$	7 is less than or equal to 7
(e) $5 \geq 1$	5 is greater than or equal to 1
(f) $12 \geq 12$	12 is greater than or equal to 12

Notice from examples (**d**) and (**f**) that $\leq$ and $\geq$ allow the numbers to be equal. The statement $a \leq b$ is actually *two* statements:

1. $a < b$, or

2. $a = b$

These statements are called **inequalities.**

Let us now look at the symbols for the algebraic **operations.**

ALGEBRAIC SYMBOLS	ENGLISH MEANING
$x + y$	Sum of x and y x plus y
$x - y$	difference of x and y x minus y
$x \cdot y$ $x(y)$ $(x)y$ $(x)(y)$ xy	product of x and y x times y
$\dfrac{x}{y}$ $x \div y$	quotient of x and y x divided by y
x^2	x squared x times x
x^3	x cubed x times x times x

Algebra

EXAMPLE 3   The following are examples of the operations.

(a) $4 + 8 = 12$

(b) $10 - 7 = 3$

(c) $3 \cdot 5 = 15$

(d) $6(10) = 60$

(e) $20 \div 4 = 5$

(f) $\dfrac{36}{9} = 4$

(g) $5^2 = 5 \cdot 5 = 25$

(h) $3^2 = 3 \cdot 3 = 9$

(i) $2^3 = 2 \cdot 2 \cdot 2 = 8$

(j) $4^3 = 4 \cdot 4 \cdot 4 = 64$

EXAMPLE 4   The following expressions are translations between algebraic symbols and English phrases.

	ALGEBRA	ENGLISH
(a)	$x + 4$	sum of x and 4
(b)	$10 - a$	difference of 10 and a
(c)	$5t$	product of 5 and t
(d)	$\dfrac{6}{z}$	quotient of 6 and z
(e)	m^2	square of m (or m squared)
(f)	p^3	cube of p (or p cubed)
(g)	$3a^2$	product of 3 and a squared
(h)	$\dfrac{b^3}{4}$	quotient of b cubed and 4
(i)	$6 + 7x$	6 plus the product of 7 and x

The operations x^2 and x^3 are examples of **exponents,** which we discuss more fully in Chapter 3. They get their special names from the square and cube of geometry. The area of a square with side x is $x \cdot x$ or x^2. The volume of a cube with side x is $x \cdot x \cdot x$ or x^3. Thus, we call these x **squared** and x **cubed.**

In English, we use punctuation to make expressions more exact. For instance,

Judy said Ethel is tacky

might be taken in two ways:

(a) Judy said, "Ethel is tacky."

(b) "Judy," said Ethel, "is tacky."

It makes a difference how the sentence is punctuated. Similarly,

$$20 - 5 - 3$$

might have two meanings:

(a) $(20 - 5) - 3 = 15 - 3 = 12$
(b) $20 - (5 - 3) = 20 - 2 = 18$

The parentheses act as quotation marks do in English. In addition to the operation symbols ($+$, $-$, $\times$, and $\div$), we have grouping symbols, such as parentheses ( ), brackets [ ], and the fraction bar ——. The symbols are used to make an expression exact, so that there are not two possible meanings.

Mathematicians have agreed on a standard order in which all operations will be done. This order is: Parentheses, Exponents, Multiplication, Division, Addition, Subtraction. It can be nicely remembered by the phrase, "Please Excuse My Dear Aunt Sally (**PEMDAS**)."

The standard order of operation is:

1. *Parentheses* (or other grouping symbols). Do all operations within parentheses or brackets, or above and below fraction bars.
2. *Exponents.* Do all squaring and cubing from left to right.
3. *Multiplication and division.* Multiply and divide from left to right.
4. *Addition and subtraction.* Add and subtract from left to right.

EXAMPLE 5 $\quad$ Simplify $2 + 3 \cdot 4$.

SOLUTION $\quad$ There are no parentheses or exponents, so we do the multiplication first, then the addition.

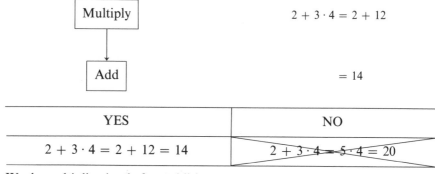

| Multiply | $2 + 3 \cdot 4 = 2 + 12$ |

| Add | $= 14$ |

YES	NO
$2 + 3 \cdot 4 = 2 + 12 = 14$	$2 + 3 \cdot 4 = 5 \cdot 4 = 20$

We do multiplication before addition.

EXAMPLE 6   Simplify $2(5 - 2)^2$.

SOLUTION   We follow the standard order: parentheses, exponent, and multiplication.

$$2(5 - 2)^2 = 2 \cdot 3^2$$

$$= 2 \cdot 9$$

$$= 18$$

EXAMPLE 7   Simplify $\dfrac{5^2 - 4}{2^3 - 1}$.

SOLUTION   Here we have a fraction bar, which gets top priority. In other words, we simplify the expressions above and below the fraction bar first, before doing the division.

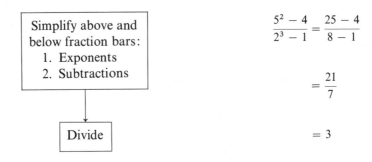

$$\frac{5^2 - 4}{2^3 - 1} = \frac{25 - 4}{8 - 1}$$

$$= \frac{21}{7}$$

$$= 3$$

EXAMPLE 8   Simplify $3 + 2 \cdot [7 - (5 - 3)^2]$.

SOLUTION   Here we have parentheses within brackets. In this case work out from the inside parentheses to the outside brackets.

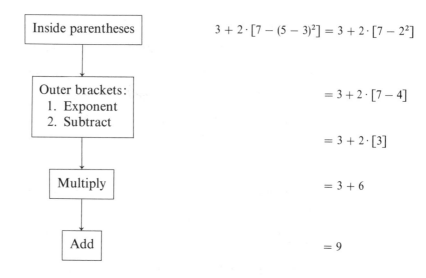

$$3 + 2 \cdot [7 - (5 - 3)^2] = 3 + 2 \cdot [7 - 2^2]$$

$$= 3 + 2 \cdot [7 - 4]$$

$$= 3 + 2 \cdot [3]$$

$$= 3 + 6$$

$$= 9$$

PROBLEM SET 1.1

State whether each statement is true or false.

See Examples 1 and 2

1. $40 = 40$ **2.** $2 + 2 = 5$ **3.** $8 \neq 7 + 1$

4. $2 \cdot 3 \neq 7$ **5.** $6 > 5$ **6.** $8 < 5$

7. $10 < 10$ **8.** $8 \geq 8$ **9.** $3 > 10$

10. $5 > 4$ **11.** $8 \leq 10$ **12.** $14 \leq 14$

Perform the indicated operations.

See Example 3

13. $2 + 7$ **14.** $3 + 11$ **15.** $10 - 3$

16. $9 - 5$ **17.** $2(7)$ **18.** $6 \cdot 5$

19. $10 \div 2$ **20.** $\dfrac{27}{9}$ **21.** 3^2

22. 10^2 **23.** 3^3 **24.** 5^3

Translate each English phrase or sentence into algebraic symbols.

See Example 1

25. a is the same number as b **26.** p is the same number as 10

27. x is not the same number as 5   **28.** t is not the same number as s

See
Example 2
29. m is less than n

30. k is less than or equal to 5

31. u is greater than or equal to v

32. 9 is greater than r

See
Example 4
33. the sum of x and 10

34. r plus s

35. the difference between a and b

36. 14 minus t

37. the product of h and k

38. 3 times n

39. x multiplied by y

40. a divided by 2

41. the quotient of p and q

42. the square of x

43. u squared

44. k cubed

45. the cube of a

46. x is the product of y and z

47. a is the sum of b and 4

48. k is m divided by 2

Simplify each of the following to a single number.

See
Example 5
49. $3 + 5 \cdot 2$

50. $4 + 3 \cdot 5$

51. $10 - 2 \cdot 3$

52. $12 - 6 \cdot 2$

53. $2 + \dfrac{8}{4}$

54. $5 - \dfrac{9}{3}$

See
Example 6
55. $2(5 - 3)$

56. $5(6 + 1)$

57. $3(2^2 + 1)$

58. $4(10 - 2^3)$

59. $4(5 - 1)^2$

60. $3(2 + 3)^2$

See
Example 7
61. $\dfrac{4 + 2}{3 - 1}$

62. $\dfrac{10 + 4}{3 + 4}$

63. $\dfrac{3 \cdot 6}{7 + 2}$

64. $\dfrac{20 + 4}{2 \cdot 3}$

65. $\dfrac{3^2 + 1}{7 - 2}$

66. $\dfrac{21 + 4}{2^3 - 3}$

See
Example 8
67. $2[3 + (2 + 2)^2]$

68. $3[10 - (4 - 1)^2]$

69. $4[(1 + 2)^3 - 20]$

70. $3[(3 - 1)^2 + 2]$

71. $\dfrac{3 - (7 - 6)}{4 - (6 - 4)}$

72. $\dfrac{10 - (2 + 2)}{8 - (4 + 1)}$

Business
applications
73. Compute the total cost (TC) to produce 100 widgets in a certain factory, which is given by

$$TC = 100^2 + 2000(100) + 200,000$$

74. Compute the balance (B) of an investment ($5000 at 7% for 2 years) given by

$$B = 5000(1.07)^2$$

Life science
application
75. Compute the likelihood (L) of at least one out of three people contracting a certain disease. This is given by

$$L = 1 - (0.8)^3$$

1.2 SUBSTITUTION, EQUALITY, AND INEQUALITY

In Section 1.1 we began to work with letters as well as specific numbers. A **variable** is a letter (or symbol) that stands for a number. The numbers that the letter might be are called the **replacement set**. (A *set* is any collection: numbers, objects, people, and so on.)

For example, if the variable x has a replacement set of $\{2, 3, 4\}$, this means that it is possible that $x = 2$, $x = 3$, or $x = 4$. The reason it is called a replacement set is that it is the set of numbers that may replace or substitute in an algebraic expression.

> **PROPERTY 1** (*Substitution Property*) If $a = b$, then a can replace b, and vice versa, in any expression without changing its value. Also, a can replace b, and vice versa, in any statement without changing its truth (or falsity).

For instance, if $x = 3$, then everywhere that x appears, we can replace it by 3. We **evaluate** an expression by replacing the variable by a specific number.

EXAMPLE 9   Evaluate $2x + 4$ when $x = 5$.

SOLUTION   We replace x by 5 and simplify.

$$2x + 4 = 2(5) + 4$$

$$= 10 + 4 = 14$$

EXAMPLE 10   Evaluate $a^2 + 5a - 6$ when $a = 2$.

SOLUTION   We replace the variable a by 2 and simplify. We use the standard order of operations: exponent, multiply, add, and subtract.

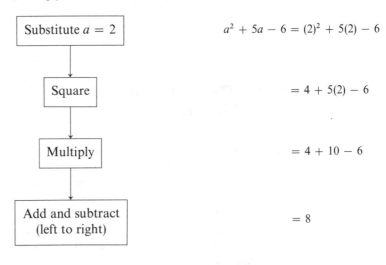

$$a^2 + 5a - 6 = (2)^2 + 5(2) - 6$$

$$= 4 + 5(2) - 6$$

$$= 4 + 10 - 6$$

$$= 8$$

EXAMPLE 11   Evaluate $\dfrac{3x - 5y}{y}$ for $x = 6$ and $y = 2$.

SOLUTION   Here we have two variables, x and y. We replace the x variable by 6 and both of the y variables by 2.

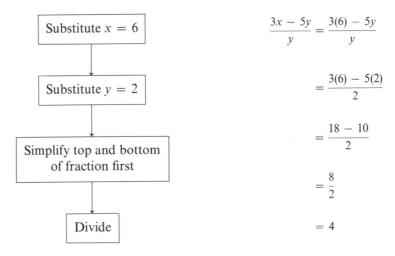

$$\frac{3x - 5y}{y} = \frac{3(6) - 5y}{y}$$

$$= \frac{3(6) - 5(2)}{2}$$

$$= \frac{18 - 10}{2}$$

$$= \frac{8}{2}$$

$$= 4$$

The substitution property can also be used to find the members of the replacement set that make a statement true.

EXAMPLE 12   Which members (if any) of the set $\{3, 4, 5\}$ will make the equation $2x + 1 = 11$ true?

SOLUTION   We substitute each number, 3, 4, and 5, for x. We then check the truth of the equation.

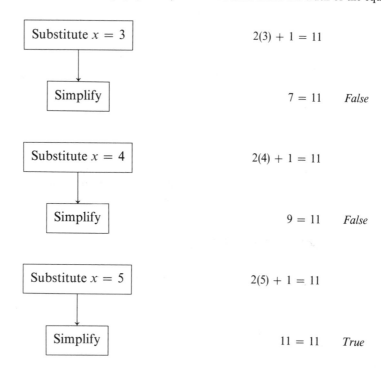

Substitute $x = 3$ $2(3) + 1 = 11$

Simplify $7 = 11$ *False*

Substitute $x = 4$ $2(4) + 1 = 11$

Simplify $9 = 11$ *False*

Substitute $x = 5$ $2(5) + 1 = 11$

Simplify $11 = 11$ *True*

The number 5 is the only number in the set that makes $2x + 1 = 11$ true. The values of a variable that make an equation true are called **solutions.** Here the solution is 5.

EXAMPLE 13   Which members (if any) of the set $\{6, 7, 8\}$ make the statement $10 - x \geq 3$ true?

SOLUTION   We substitute each number, 6, 7, and 8, into the inequality. Then we check to see if it is true or false.

Substitute $x = 6$ $10 - 6 \geq 3$

Simplify $4 \geq 3$ *True*

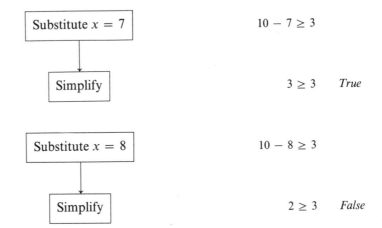

The numbers from the set that make this inequality true are 6 and 7. (There are other numbers in addition to 6 and 7 that also make the inequality true.)

PROBLEM SET 1.2

Evaluate the following expressions.

See Example 9

1. $x + 3$ for $x = 2$

2. $x - 7$ for $x = 10$

3. $5a$ for $a = 4$

4. $7t$ for $t = 8$

5. $\dfrac{x}{6}$ for $x = 24$

6. $\dfrac{40}{m}$ for $m = 5$

7. $2u + 5$ for $u = 2$

8. $5k - 3$ for $k = 4$

9. $12 - 3x$ for $x = 1$

10. $20 - 6y$ for $y = 2$

See Example 10

11. x^2 for $x = 6$

12. a^3 for $a = 2$

13. $t^2 + t$ for $t = 2$

14. $u^3 + 3$ for $u = 1$

15. $x^2 - 3x + 1$ for $x = 5$

16. $m^3 + m + 1$ for $m = 2$

17. $\dfrac{x + 3}{x - 3}$ for $x = 6$

18. $\dfrac{k + 7}{k - 3}$ for $k = 5$

19. $\dfrac{3y + 8}{y}$ for $y = 2$

20. $\dfrac{2s + 7}{s - 7}$ for $s = 10$

See Example 11

21. $x + y$ for $x = 3$ and $y = 5$

22. $x - y$ for $x = 7$ and $y = 1$

23. $2a + 3b$ for $a = 5$ and $b = 3$

24. $4m - 3n$ for $m = 4$ and $n = 2$

25. $x^2 + y^2$ for $x = 2$ and $y = 5$

26. $a^3 - b^3$ for $a = 3$ and $b = 2$

27. $\dfrac{x + y}{x - y}$ for $x = 5$ and $y = 4$

28. $\dfrac{2u - v}{u + v}$ for $u = 2$ and $v = 1$

For each of the following equations or inequalities, decide which elements of the given replacement sets make the equation or inequality true. In other words, find the solutions that are in the replacement set.

See
Example 12

29. $x + 5 = 9$ $\{3, 4, 5\}$ **30.** $a - 6 = 3$ $\{7, 8, 9\}$

31. $2k - 3 = 7$ $\{5, 6, 7\}$ **32.** $3x + 4 = 22$ $\{5, 6, 7\}$

33. $x^2 + 1 = 10$ $\{2, 3, 4\}$ **34.** $t^3 - 2 = 6$ $\{2, 3, 4\}$

35. $x^2 - 4 = (x - 2)(x + 2)$ $\{3, 4, 5\}$

36. $r^2 - 9 = (r + 3)(r - 3)$ $\{5, 6, 7\}$

See
Example 13

37. $x + 5 \le 8$ $\{2, 3, 4\}$ **38.** $a - 7 \ge 2$ $\{7, 8, 9\}$

39. $2x > 7$ $\{3, 4, 5\}$ **40.** $3t < 8$ $\{1, 2, 3\}$

41. $2x + 1 \ge 3$ $\{1, 2, 3\}$ **42.** $3r - 2 \le 10$ $\{2, 3, 4\}$

43. $x^2 + 2 \ge 1$ $\{1, 2, 3\}$ **44.** $k^2 < 1$ $\{4, 5, 6\}$

Science
applications

45. The relation between Celsius temperature C and Fahrenheit temperature F is given by

$$C = \tfrac{5}{9}(F - 32)$$

Compute C if
(a) $F = 50$ degrees
(b) $F = 77$ degrees

46. The distance d that an object falls in t seconds is given by

$$d = 16t^2$$

Compute d if
(a) $t = 2$ seconds
(b) $t = 3$ seconds

Health application   **47.**   Young's formula for a child's dosage C of a medicine is given by

$$C = \left(\frac{A}{A + 12}\right)D$$

where A is the child's age and D is the adult's dosage.   Compute C if $A = 6$ years and $D = 300$ milligrams.

Business application   **48.**   Straight-line annual depreciation D of a machine is given by

$$D = \frac{C - S}{L}$$

where C is the original cost, S is the salvage value, and L is the life of the machine.   Compute D if $C = \$50{,}000$, $S = \$5000$, and $L = 10$ years.

1.3   THE NUMBER LINE AND THE REAL NUMBERS

The most convenient way to picture numbers is with a **number line.**

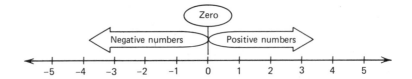

The line continues without end in both directions.   The point 0 is called the **origin.** Associated with each point of the line is a number (called its **coordinate**), and the set of all these numbers is called the **real numbers.**

The points to the right of 0 are the **positive real numbers.** These are written with a positive sign $(+)$ or, more often, no sign at all.   For instance, $+4$ (or just 4) is a positive number.

The points to the left of 0 are called the **negative real numbers.** These are written with a negative sign $(-)$.   For instance, -3 is a negative number.

EXAMPLE 14   Give the coordinates of the points on the number line.

SOLUTION   In some cases we have to estimate the coordinates.

$$A = 2 \qquad B = 3\tfrac{1}{2} \qquad C = 0 \qquad D = -1 \qquad E = -2\tfrac{1}{4}$$

Although most students are familiar with the positive numbers from arithmetic, the negative numbers may be a new idea for many students. Negative numbers are most commonly used to represent opposites for positive quantities.

EXAMPLE 15   The following table shows how negative numbers are used to contrast with positive quantities.

POSITIVES		NEGATIVES	
PHRASE	NUMBER	PHRASE	NUMBER
$100 in assets	+100	$100 in debt	−100
3000 feet above sea level	+3000	3000 feet below sea level	−3000
20 seconds after blast-off	+20	20 seconds before blast-off	−20
5-yard gain	+5	5-yard loss	−5
10°C above freezing	+10	10°C below freezing	−10

Notice that the positive and negative numbers come in pairs, such as 5 and −5, or 100 and −100.

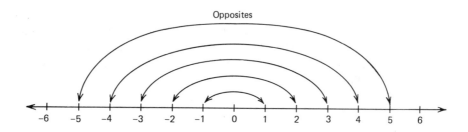

Opposites

The illustration shows that every number (except 0) has an **opposite** on the other side of 0. We use the notation $-x$ to stand for the opposite of x. (This is also called the **additive inverse.**)

EXAMPLE 16   The following are examples of numbers and their opposites.

NUMBER (x)	OPPOSITE $(-x)$
8	-8
2	-2
-6	6
-5.2	5.2
0	0
$\dfrac{1}{2}$	$-\dfrac{1}{2}$
$-\pi$	π

Notice that 0 is its own opposite.

Is $-x$ always a negative number? Let us see.

1. Suppose that $x = 7$; then $-x = -7$, which is negative.
2. Suppose that $x = -4$; then $-x$ is the opposite of -4. This is 4. In other words, $-x = -(-4) = 4$, which is positive.

In general, we have the following property.

PROPERTY 2   (*Double Negative Property*)   For any real number x,

$$-(-x) = x$$

Notice that each pair of opposites, such as 4 and -4, are the same distance, 4, from 0. This distance from the origin is always zero or positive, and is called the **absolute value.** We write the absolute value of x as $|x|$. See the illustration.

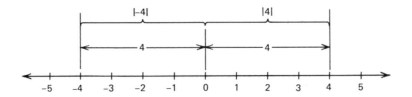

EXAMPLE 17   The following are examples of absolute value, the distance from 0.

(a) $|7| = 7$

(b) $|-8| = 8$

(c) $|14| = 14$

(d) $|0| = 0$

(e) $-|-2| = -2$ (since $|-2| = 2$)

We can use the number line to help visualize inequalities.

ALGEBRA	ENGLISH	NUMBER LINE
$a < b$	*a* is *less* than *b*	*a* is to the *left* of *b*
$a = b$	*a* is *equal* to *b*	*a* is the *same* point as *b*
$a > b$	*a* is *greater* than *b*	*a* is to the *right* of *b*

EXAMPLE 18   A number line helps us picture the following inequalities.

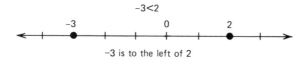

-3 is to the left of 2

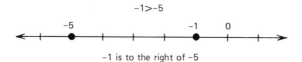

-1 is to the right of -5

PROBLEM SET 1.3

Estimate the coordinates of the points on the number line.

See
Example 14

1. Point *A*   **2.** Point *B*   **3.** Point *C*

4. Point *D*   **5.** Point *E*   **6.** Point *F*

On a number line, locate the following points.

See Example 14

7. $A = 2$ **8.** $B = 1\frac{1}{2}$ **9.** $C = 3\frac{1}{3}$

10. $D = -1$ **11.** $E = -3.7$ **12.** $F = 2\frac{1}{4}$

Replace the following phrases by an appropriate positive or negative number.

See Example 15

13. 55 miles per hour forward

14. A $15 loss at poker

15. A $20 price increase

16. A 6-pound weight gain

17. Running a movie backward for 4 minutes

18. A $2\frac{1}{4}$-point drop in a stock price

19. A 200-Btu (British thermal unit) heat loss

20. 20°C above freezing

21. A $4000 profit from a house sale

22. A $200 debt

Complete the following table.

See Examples 16 and 17

	NUMBER	OPPOSITE	ABSOLUTE VALUE
23.	18		
24.	−3		
25.	−11		
26.	12		
27.		−2	
28.		10	
29.		9	
30.		−7	

Write the following numbers as simply as possible.

See Examples 16 and 17

31. $-(-6)$ **32.** $-(-7)$ **33.** $|12|$

34. $|-3|$ **35.** $-|6|$ **36.** $-|-8|$

Label the following statements as true or false.

See
Example 18

37. $2 < -3$ **38.** $-4 < -3$ **39.** $-3 > 0$

40. $-3 > -5$ **41.** $-2 \geq -1$ **42.** $-6 \leq -8$

Business application

43. Locate the following profits on the number line.

$$P_1 = \$10{,}000 \qquad P_2 = -\$15{,}000 \qquad P_3 = \$25{,}000 \qquad P_4 = -\$3000$$

Psychology application

44. Locate the following moods ($-3 = $ very depressed; $0 = $ neutral; $+3 = $ very happy) on a number line.

$$A(\text{Alice}) = +2 \qquad B(\text{Bob}) = -3 \qquad C(\text{Carol}) = 0 \qquad D(\text{Dave}) = 1$$

1.4   ADDITION OF REAL NUMBERS

We can use the real-number line to help us picture the addition of real numbers.

EXAMPLE 19   Add $4 + 3$.

SOLUTION   This may be a first-grade problem, but it is still a good place to start. These are both positive numbers, so we can imagine them as assets. If we have \$4 in assets and \$3 in assets, together we have \$7 in assets. On a number line, this appears as follows.

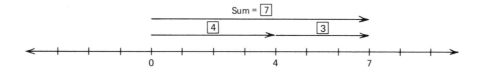

We place the 3-arrow after the 4-arrow, and get a 7-arrow. Thus, $4 + 3 = 7$.

EXAMPLE 20   Add $-2 + (-3)$.

SOLUTION   This time the numbers are both negative, so we can imagine them as debts. If we have a $2 debt and a $3 debt, together we have a $5 debt.

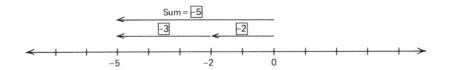

On the number line we see that a -3-arrow and a -2-arrow produce a -5-arrow. Thus, $-2 + (-3) = -5$.

These two examples suggest the following rule.

To add two real numbers with the *same* sign:

1. Add their absolute values.
2. Give the answer the common sign.

EXAMPLE 21   The following additions are of real numbers with the same sign. Notice that if both numbers are positive, the sum is positive. Similarly, if both numbers are negative, the sum is negative.

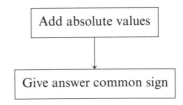

(a) $5 + 7 = 12$

(b) $(+3) + (+11) = +14$

(c) $-4 + (-9) = -13$

(d) $-10 + (-2) = -12$

(e) $-20 + (-30) = -50$

EXAMPLE 22   Add $7 + (-2)$.

SOLUTION   These numbers have the opposite signs. Suppose that we have a $7 asset and a $2 debt; then our net is $5 in assets.

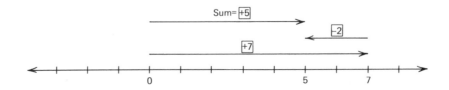

On the number line, if we attach a -2-arrow to the end of a 7-arrow, we end up with a 5-arrow. Thus, $7 + (-2) = 5$.

EXAMPLE 23   Add $-6 + 4$.

SOLUTION   Again, we have opposite signs. Suppose that we have a $6 debt and $4 assets; then together we have a net debt of $2.

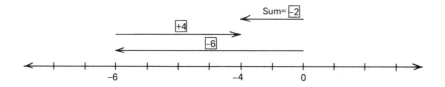

On the number line, we see that attaching a 4-arrow to a -6-arrow gives a -2-arrow; thus, $-6 + 4 = -2$.

These examples suggest another rule.

To add two real numbers of *opposite* signs:

1. Subtract the absolute values (larger − smaller).
2. Give the answer the sign of the number with the larger absolute value.

EXAMPLE 24   The following additions are of real numbers with opposite signs. Notice that the sum always has the sign of the number with the larger absolute value.

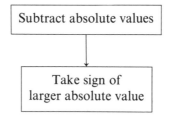

(a) $2 + (-9) = -7$

(b) $6 + (-1) = 5$

(c) $-7 + 4 = -3$

(d) $-2 + 10 = 8$

(e) $-20 + 15 = -5$

EXAMPLE 25   Add $-3 + (-7 + 9) + (-8)$.

SOLUTION   Recall that we must perform the operation within the parentheses first. Then we do additions, left to right.

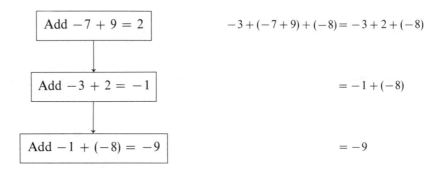

$$-3 + (-7 + 9) + (-8) = -3 + 2 + (-8)$$

$$= -1 + (-8)$$

$$= -9$$

PROBLEM SET 1.4

Display the following additions on a number line.

See
Examples
19, 20, 22, and 23

1. $2 + 5$ **2.** $-3 + (-4)$ **3.** $-2 + (+1)$

4. $-1 + 5$ **5.** $-2 + 5$ **6.** $3 + (-8)$

Find the following sums.

See
Examples
21 and 24

7. $2 + (-9)$ **8.** $11 + 3$ **9.** $14 + (-5)$

10. $-10 + (-6)$ **11.** $-6 + 15$ **12.** $-8 + 7$

13. $9 + (-10)$ **14.** $8 + (-5)$ **15.** $12 + (+6)$

16. $-4 + 13$ **17.** $-7 + (-9)$ **18.** $14 + (-11)$

19. $-20 + (-15)$ **20.** $30 + (-17)$ **21.** $20 + (+16)$

22. $17 + (-19)$ **23.** $-15 + (-3)$ **24.** $-17 + 13$

See
Example 25

25. $2 + 3 + 6$ **26.** $2 + 5 + (-8)$

27. $5 + (-2) + (-6)$ **28.** $-4 + (-2) + 9$

29. $-6 + (-9) + (-3)$ **30.** $-2 + (-5) + (-1)$

31. $-4 + 3 + (-8)$ **32.** $-5 + 10 + (-3)$

33. $-6 + 11 + 2$ **34.** $-8 + 2 + 3$

35. $12 + (-6) + 1$ **36.** $2 + (-5) + 1$

37. $-2 + (-6 + 8) + 3$ **38.** $-4 + 7 + (-8 + 10)$

39. $4 + (-2 + 5) + (-3)$ **40.** $-6 + [3 + (-8)] + (-4)$

41. $[-5 + (-4)] + [-3 + 8]$ **42.** $[3 + (-4)] + [-2 + (-5)]$

43. $-1 + [-8 + 3] + [2 + (-10)] + [-4 + (-3)]$

44. $5 + [-2 + (-9)] + [-3 + 12] + [4 + (-6)]$

Evaluate the following expressions.

*See
Examples
9, 11, 21, and 24*

45. $x + 7$ for $x = -8$ **46.** $-3 + a$ for $a = 9$

47. $-8 + t$ for $t = -5$ **48.** $u + (-6)$ for $u = 1$

49. $x + y$ for $x = -3$ and $y = -5$

50. $a + b$ for $a = 7$ and $b = -11$

*Health
application*

51. A patient's weekly weight gains (and losses) are recorded as

$$+2, \quad -1, \quad -1.5, \quad +1, \quad -2.5, \quad -0.5, \quad +1, \quad -1.5$$

What is the patient's net gain (or loss)?

*Business
application*

52. A business lists its assets $(+)$ and debts $(-)$ as

$$\$10,000, \quad -\$7000, \quad \$2500, \quad \$3000, \quad -\$4500$$

Find the company's net worth.

*Ecology
application*

53. Over the period of 1 week, the daily changes in the air pollution (from lead) from a factory were (in micrograms per cubic meter)

$$+3, \quad -10, \quad +5, \quad +6, \quad -2, \quad -11, \quad -13$$

Find the net change over the week.

*Chemical
application*

54. In chemistry, elements and compounds have valences, which add as real numbers do. Find the net valences of
(a) Cr_2O_7 if the valence of Cr_2 is $+12$ and the valence of O_7 is -14.
(b) Fe_2O_3 if the valence of Fe_2 is $+6$ and the valence of O_3 is -6.

1.5   SUBTRACTION OF REAL NUMBERS

We now show how subtraction is the opposite of addition.

EXAMPLE 26   Subtract $7 - 4$.

SOLUTION   Consider this as a temperature problem: If it is 4° one hour and 7° the next hour, what is the temperature jump?

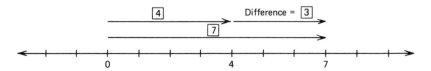

We see on the number line that the difference between 7° and 4° is 3°. This is also the result that we get by adding $7 + (-4)$.

EXAMPLE 27   Subtract $5 - (-3)$.

SOLUTION   Consider this as a temperature problem: If it is $-3°$ one hour and 5° the next hour, what is the temperature jump?

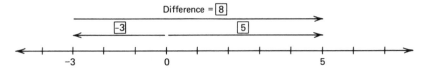

We see on the number line that the difference is 8°. Thus, $5 - (-3) = 8$. Also, this is the same result as adding $5 + (+3) = 8$.

These examples suggest the next rule.

> To subtract real numbers, add the opposite of the second number.

This is written as follows:

> **PROPERTY 3**   For any real numbers x and y,
>
> $$x - y = x + (-y)$$

Another way of stating the subtraction rule is that we *change the sign of the second number and add.*

EXAMPLE 28   The following subtractions use the fact that subtraction means *add the opposite of the second number.*

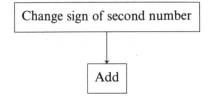

(a) $7 - 3 = 7 + (-3) = 4$

(b) $5 - 13 = 5 + (-13) = -8$

(c) $-3 - 9 = -3 + (-9) = -12$

(d) $-5 - (-11) = -5 + (+11) = 6$

(e) $-7 - (-1) = -7 + (+1) = -6$

EXAMPLE 29   Evaluate $15 - (x - y)$ for $x = -2$ and $y = 3$.

SOLUTION   We first substitute and then do the subtraction within the parentheses.

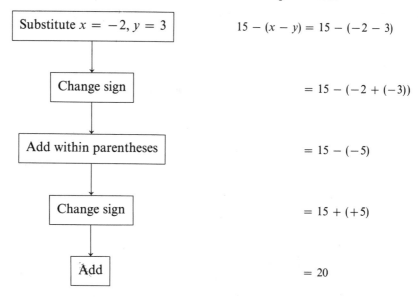

$$15 - (x - y) = 15 - (-2 - 3)$$

$$= 15 - (-2 + (-3))$$

$$= 15 - (-5)$$

$$= 15 + (+5)$$

$$= 20$$

EXAMPLE 30   Simplify $T = 3 + (-5) - 7 - (-8) + (-4)$.

SOLUTION   We start by changing all subtractions to additions of opposites.

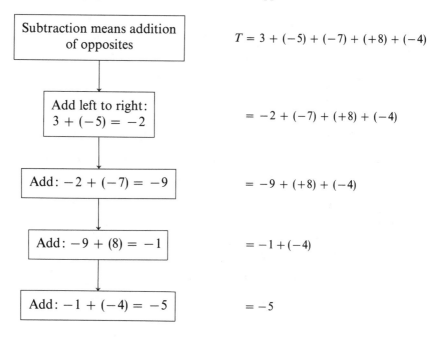

$$T = 3 + (-5) + (-7) + (+8) + (-4)$$

$$= -2 + (-7) + (+8) + (-4)$$

$$= -9 + (+8) + (-4)$$

$$= -1 + (-4)$$

$$= -5$$

Another way to add a sequence of positive and negative numbers is to group all the positives and all the negatives.

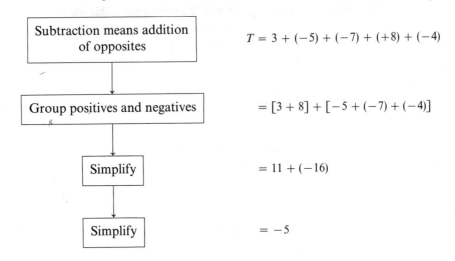

$$T = 3 + (-5) + (-7) + (+8) + (-4)$$

$$= [3 + 8] + [-5 + (-7) + (-4)]$$

$$= 11 + (-16)$$

$$= -5$$

PROBLEM SET 1.5

Display the following subtractions on a number line.

See Examples 26 and 27

1. $2 - 7$ **2.** $3 - 4$ **3.** $-1 - 3$

4. $-3 - 6$ **5.** $3 - (-5)$ **6.** $2 - (-7)$

Simplify the following expression.

See Example 28

7. $3 - 10$ **8.** $5 - 7$ **9.** $-3 - 5$

10. $10 - (-12)$ **11.** $8 - 3$ **12.** $4 - (-7)$

13. $-9 - (-3)$ **14.** $-6 - 7$ **15.** $-5 - (-8)$

16. $10 - 18$ **17.** $-4 - (-9)$ **18.** $9 - (-2)$

19. $3 - 9$ **20.** $-5 - (-2)$ **21.** $-7 - 10$

22. $-11 - 13$ **23.** $-9 - (-15)$ **24.** $8 - (-21)$

See Example 29

25. $3 - (8 - 1)$ **26.** $4 - (7 - 9)$

27. $-4 - (7 - 3)$ **28.** $10 - (2 - 6)$

29. $-8 - [3 - (-5)]$ **30.** $5 - [-8 - (-7)]$

31. $-10 - [-6 - (-8)]$ **32.** $-5 - [9 - (-8)]$

33. $-11 - [-3 - 7]$ **34.** $-12 - [-5 - (-6)]$

See
Example 30

35. $-7 - 5 + 2 - 3 - (-6)$ **36.** $-8 - (-7) - 6 + 4 - 2$

37. $-3 - (-6) + 10 - 4 - (-3)$ **38.** $10 + (-4) - 5 - (-2) + 3$

39. $7 - 9 + (-3) - (-5) - 10$ **40.** $-1 - (-5) - 6 - 8 + (-4)$

41. $-10 - [-2 + 5] - [5 - (-7)] - [8 - 11]$

42. $8 - [3 - 5] - [1 + (-6)] - [-2 - (-7)]$

Evaluate the following expressions.

See
Examples
9, 11, and 29

43. $x - 4$ for $x = 3$ **44.** $6 - a$ for $a = -5$

45. $-1 - t$ for $t = -7$ **46.** $-3 - r$ for $r = 6$

47. $a - b$ for $a = 7$ and $b = -5$

48. $x - y$ for $x = -5$ and $y = -4$

49. $2 - u - v$ for $u = 3$ and $v = -7$

50. $p - q - 8$ for $p = -5$ and $q = -10$

Business
application

51. In business, we have the formula

change in profit = change in revenue − change in cost

Use this formula to complete the following table.

CHANGE IN PROFIT	CHANGE IN REVENUE	CHANGE IN COST
?	\$ 20,000	\$ 15,000
?	30,000	38,000
?	−12,000	−17,000
?	−8,000	2,000

Science
application

52. The change of temperature in a substance is given by

temperature change = final temperature − original temperature

Use this formula to complete the following table.

TEMPERATURE CHANGE	FINAL TEMPERATURE	ORIGINAL TEMPERATURE
?	18°C	2°C
?	−14°C	27°C
?	−2°C	−9°C
?	5°C	−13°C

1.6   MULTIPLICATION OF REAL NUMBERS

Now, we look at (but do not prove) the rules for multiplying real numbers. Recall from arithmetic that multiplying by a whole number can be considered as *repeated addition*. Similarly, we can consider multiplying by a negative integer as *repeated subtraction*. The following table shows all the possibilities for multiplying positive and negative numbers.

Factors	Meaning	Product
$3 \cdot 4$	Add 4 three times $= 4 + 4 + 4 = 12$	12
$3 \cdot (-4)$	Add (-4) three times $= (-4) + (-4) + (-4) = -12$	-12
$(-3) \cdot 4$	Subtract 4 three times $= -4 - 4 - 4$ $= (-4) + (-4) + (-4) = -12$	-12
$(-3) \cdot (-4)$	Subtract (-4) three times $= -(-4) - (-4) - (-4)$ $= 4 + 4 + 4 = 12$	12

Notice that all the products have an absolute value of 12; only the signs are different. When we multiplied two positives or two negatives, we got a *positive* number. When we multiplied one positive and one negative, we got a *negative* product. This suggests the following general rule.

> To multiply two real numbers:
>
> **1.** Multiply their absolute values.
> **2a.** If the numbers have the *same* sign, the product is *positive*.
> **2b.** If the numbers have *opposite* signs, the product is *negative*.

EXAMPLE 31   The following multiplications are pairs of real numbers with the *same* sign. The products are all *positive*.

(a) $2 \cdot 5 = 10$

(b) $3 \cdot 7 = 21$

(c) $(-4)(-6) = 24$

(d) $(-10)(-3) = 30$

(e) $\left(-\dfrac{1}{2}\right)\left(-\dfrac{1}{3}\right) = \dfrac{1}{6}$

EXAMPLE 32   The following multiplications are pairs of real numbers with *opposite* signs. The products are all *negative*.

(a) $(-5)(7) = -35$

(b) $(-4)(10) = -40$

(c) $(6)(-3) = -18$

(d) $(8)(-2) = -16$

(e) $\left(-\dfrac{2}{3}\right)\left(\dfrac{5}{7}\right) = -\dfrac{10}{21}$

Let us now state some simple, but important, properties for multiplication.

PROPERTY 4   For any real number x,

(a) $1 \cdot x = x \cdot 1 = x$

(b) $0 \cdot x = x \cdot 0 = 0$

(c) $(-1)x = x(-1) = -x$

EXAMPLE 33   The following examples show the three properties stated above.

(a) $1 \cdot 7 = 7$

(b) $(-4) \cdot 1 = -4$

(c) $0 \cdot (-5) = 0$

(d) $15 \cdot 0 = 0$

(e) $(-1) \cdot \dfrac{1}{2} = -\dfrac{1}{2}$

(f) $(-6)(-1) = 6$

EXAMPLE 34   Simplify $T = -3(-2 - 5) - 7(-4 + 6)$.

$$-3(-7) \; -7(o)$$
$$21 + -14 = 7$$

SOLUTION   We have to follow the standard order of operation: parentheses, multiplication, addition, and subtraction. First, we change all subtractions to additions of opposites.

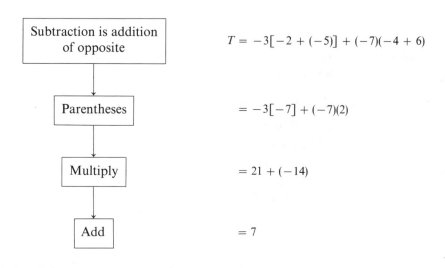

Subtraction is addition of opposite	$T = -3[-2 + (-5)] + (-7)(-4 + 6)$
Parentheses	$= -3[-7] + (-7)(2)$
Multiply	$= 21 + (-14)$
Add	$= 7$

EXAMPLE 35   Evaluate $(x - 7)(y + 9)$ for $x = 3$ and $y = -2$.

$$(3-7)(-2+9)$$
$$-4 \cdot -7 = 28$$

SOLUTION   We first substitute $x = 3$ and $y = -2$; then we simplify.

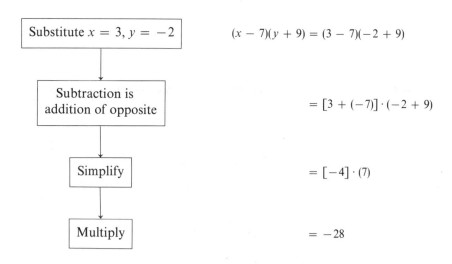

Substitute $x = 3, y = -2$	$(x - 7)(y + 9) = (3 - 7)(-2 + 9)$
Subtraction is addition of opposite	$= [3 + (-7)] \cdot (-2 + 9)$
Simplify	$= [-4] \cdot (7)$
Multiply	$= -28$

PROBLEM SET 1.6

Find the following products.

See
Examples
31 and 32

1. $2 \cdot 7$	**2.** $-2 \cdot (7)$	**3.** $4(-6)$
4. $-3(-7)$	**5.** $-4(5)$	**6.** $10(-4)$
7. $5(6)$	**8.** $6(-4)$	**9.** $-5(-8)$
10. $(-8)(-6)$	**11.** $-4(12)$	**12.** $9(-3)$
13. $-11(-4)$	**14.** $-3(-5)$	**15.** $5(10)$
16. $-7(5)$	**17.** $\dfrac{1}{3}\left(-\dfrac{1}{4}\right)$	**18.** $-\dfrac{1}{2}\left(\dfrac{1}{5}\right)$

See
Example 33

19. $6 \cdot 0$	**20.** $(1)9$	**21.** $-8(-1)$
22. $-7(1)$	**23.** $0\left(\dfrac{1}{2}\right)$	**24.** $-1(14)$

Simplify the following expressions using the standard order of operation.

See
Example 34

25. $-2 + 3 \cdot 4$ **26.** $4 - (-3)(5)$

27. $(-2)(3) - (5)(-1)$ **28.** $(-3)(4) + (-2)(-7)$

29. $-5(6 - 10)$ **30.** $2(-3 - 5)$

31. $10 - 4(2 - 5)$ **32.** $-3 + (-4)(1 - 8)$

33. $-2(1 - 6) - 4(-2 + 8)$ **34.** $5(3 - 5) - 4(-2 - 6)$

Evaluate the following expressions.

See
Example 35

35. $-7x$ for $x = 10$ **36.** $-10a$ for $a = -3$

37. $6r$ for $r = -4$ **38.** $-2t$ for $t = -8$

39. $2x - 5$ for $x = -3$ **40.** $-3a + 4$ for $a = -5$

41. $-3(x - 4)$ for $x = 2$ **42.** $-5(6 - a)$ for $a = -2$

43. $2 + 3x - 4y$ for $x = -2$ and $y = 4$

44. $3 - 2a + 3b$ for $a = 6$ and $b = -1$

45. $(u - 6)(10 - v)$ for $u = 3$ and $v = 4$

46. $(x + 1)(6 + y)$ for $x = -5$ and $y = -7$

Chemistry application

47. In chemistry, the *valence* of an ion of an element is the product of the valence of the element and the number of atoms. Complete the following table.

ELEMENT (VALENCE)	ION	NET VALENCE
Cr $(+6)$	Cr_2	$2(+6) = +12$
O (-2)	O_7	$7(-2) = -14$
H $(+1)$	H_3	?
N (-3)	N_1	?
O (-2)	O_3	?
C $(+4)$	C_2	?

Business application

48. The extra profit (P) in producing more (or less) items is the *marginal profit* (MP) times the number of extra units produced (U). Written as a formula, this is

$$P = (MP)(U)$$

Use this formula to complete the following table.

EXTRA PROFIT (P)	MARGINAL PROFIT (MP)	EXTRA UNITS (U)
?	50	100
?	-80	200
?	200	-30
?	-500	-20

1.7   DIVISION OF REAL NUMBERS

Division is the opposite of multiplication. Recall from arithmetic that

$$10 \div 2 = \boxed{?} \quad \text{means} \quad 2 \cdot \boxed{?} = 10 \quad (\text{thus, } \boxed{?} = 5)$$

Division has the same meaning with real numbers. For instance,

$$-10 \div 2 = \frac{-10}{2} = \boxed{?} \quad \text{means} \quad 2 \cdot \boxed{?} = -10$$

We can see that $\boxed{?} = -5$. Using this thinking, we can derive the rules for division that are very similar to the rules for multiplication.

To divide two real numbers:

1. Divide their absolute values.

2a. If the numbers have the *same* sign, the quotient is *positive*.

2b. If the numbers have *opposite* signs, the quotient is *negative*.

EXAMPLE 36   The following divisions are pairs of numbers with the *same* sign. The quotients are all *positive*.

(a) $\dfrac{10}{2} = 5$

(b) $\dfrac{-12}{-6} = 2$

(c) $\dfrac{-20}{-4} = 5$

(d) $\dfrac{36}{9} = 4$

(e) $\dfrac{-100}{-1} = 100$

EXAMPLE 37   The following divisions are pairs of numbers with *opposite* signs. The quotients are all *negative*.

(a) $\dfrac{-12}{2} = -6$

(b) $\dfrac{-25}{5} = -5$

(c) $\dfrac{21}{-3} = -7$

(d) $\dfrac{30}{-10} = -3$

(e) $\dfrac{18}{-1} = -18$

EXAMPLE 38   Simplify $\dfrac{(5)(-4) - (4)(-2)}{2(4 - 5)}$.

SOLUTION   We first simplify above and below the fraction bar; then we divide.

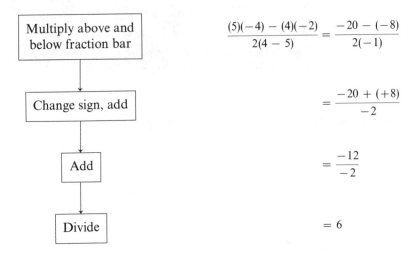

$$\frac{(5)(-4) - (4)(-2)}{2(4 - 5)} = \frac{-20 - (-8)}{2(-1)}$$

$$= \frac{-20 + (+8)}{-2}$$

$$= \frac{-12}{-2}$$

$$= 6$$

Special mention should be given to the number 0. *We can never divide by zero.*

YES	NO
$\dfrac{0}{7} = 0$	$\dfrac{7}{0} = 0$

We cannot divide by zero. The fraction $\dfrac{7}{0}$ makes no sense. Why? Suppose there is a quotient q so that $\dfrac{7}{0} = q$. Then $7 = 0 \cdot q$. But $0 \cdot q = 0$, and $7 = 0$ is false. Therefore, there is no such q.

YES	NO
$\dfrac{6}{6} = 1$	$\dfrac{0}{0} = 1$

The fraction $\dfrac{0}{0}$ makes no sense. Why? Suppose there is a quotient q so that $\dfrac{0}{0} = q$. Then $0 = 0 \cdot q$. What q makes this true? Any $q : 1, 2, 3\frac{1}{2}, \pi, 7.9, -6$, and so on may be true. Since q might be any number, we say that there is *no* quotient $\dfrac{0}{0}$. (A quotient should be only one number, not many.)

Just as subtraction can be viewed as addition of the opposite, it is often convenient to view division as *multiplication by the reciprocal*. Two numbers whose product is 1 are called **reciprocals.** (They are also called **multiplicative inverses.**)

EXAMPLE 39 The following are numbers and their reciprocals. (Their products are 1.)

	NUMBER	RECIPROCAL	
(a)	2	$\dfrac{1}{2}$	since $2 \cdot \dfrac{1}{2} = 1$
(b)	-5	$\dfrac{1}{-5}$	since $(-5)\left(\dfrac{1}{-5}\right) = 1$
(c)	$\dfrac{4}{3}$	$\dfrac{3}{4}$	since $\dfrac{4}{3} \cdot \dfrac{3}{4} = 1$
(d)	$\dfrac{-11}{7}$	$\dfrac{7}{-11}$	since $\left(\dfrac{-11}{7}\right)\left(\dfrac{7}{-11}\right) = 1$
(e)	t	$\dfrac{1}{t}$	since $t \cdot \dfrac{1}{t} = 1$
(f)	0	(none)	

Notice that 0 has no reciprocal since $\dfrac{1}{0}$ is not defined.

PROPERTY 5 For any real numbers a and b $(b \neq 0)$,

$$\frac{a}{b} = a \cdot \left(\frac{1}{b}\right)$$

This is exactly the "invert and multiply" rule learned for dividing fractions. This rule says that *division is multiplication by the reciprocal.*

EXAMPLE 40   The following divisions use the fact that division can be replaced by multiplication by the reciprocal.

Division means multiply by reciprocal

(a) $-10 \div 5 = -10 \cdot \dfrac{1}{5} = -2$

(b) $\dfrac{4}{3} \div \dfrac{-5}{7} = \dfrac{4}{3} \cdot \dfrac{-7}{5} = \dfrac{-28}{15}$

(c) $-8 \div \left(\dfrac{-1}{3}\right) = -8 \cdot (-3) = 24$

PROBLEM SET 1.7

Find the following quotients. (Be careful where zero is involved—see Yes/No boxes.)

See Examples 36 and 37

1. $\dfrac{4}{-2}$ **2.** $\dfrac{-6}{3}$ **3.** $\dfrac{-10}{-2}$

4. $\dfrac{-26}{-13}$ **5.** $\dfrac{30}{15}$ **6.** $\dfrac{18}{-6}$

7. $\dfrac{32}{-8}$ **8.** $\dfrac{-16}{-2}$ **9.** $\dfrac{-20}{-10}$

10. $\dfrac{0}{-5}$ **11.** $\dfrac{30}{-5}$ **12.** $\dfrac{14}{7}$

13. $\dfrac{14}{0}$ **14.** $\dfrac{-20}{4}$ **15.** $\dfrac{-90}{-3}$

16. $\dfrac{-42}{6}$ **17.** $\dfrac{0}{-6}$ **18.** $\dfrac{100}{-2}$

19. $\dfrac{-50}{-50}$ **20.** $\dfrac{0}{0}$

Simplify the following expressions using the standard order of operation. (Recall *PEMDAS*.)

See Example 38

21. $12 + \dfrac{-6}{3}$ **22.** $-4 - \dfrac{8}{-4}$

23. $\dfrac{3(-6)}{9}$ **24.** $\dfrac{-30}{(-2)(-5)}$

25. $\dfrac{(-10)(10)}{(-2)(-25)}$

26. $\dfrac{(-8)(-8)}{(-2)(-4)}$

27. $\dfrac{2 - 10}{1 - 3}$

28. $\dfrac{-6 + 20}{2 - 9}$

29. $\dfrac{2(-3) + 4(-1)}{1 - 6}$

30. $\dfrac{-6(-2) - 4(-5)}{-2 + 6}$

Evaluate the following expressions.

See Examples 35 and 38

31. $\dfrac{x}{6}$ for $x = -12$

32. $\dfrac{a}{-3}$ for $a = -27$

33. $\dfrac{-10}{t}$ for $t = 5$

34. $\dfrac{-20}{u}$ for $u = -5$

35. $\dfrac{x + 4}{x - 4}$ for $x = 2$

36. $\dfrac{y + 10}{y - 8}$ for $y = 5$

37. $\dfrac{a + 12}{b - 5}$ for $a = -2$ and $b = -5$

38. $\dfrac{r - 10}{s + 8}$ for $r = 4$ and $s = -2$

See Example 39

Give the reciprocals (if any) for each of the following numbers.

39. 8

40. -10

41. $\dfrac{-2}{3}$

42. $\dfrac{5}{6}$

43. 0

44. -1

Perform the following divisions using the "multiply by the reciprocal" rule.

See Example 40

45. $-14 \div \dfrac{1}{2}$

46. $\dfrac{1}{3} \div (-7)$

47. $\dfrac{2}{3} \div \dfrac{-5}{11}$

48. $\dfrac{-5}{6} \div \dfrac{2}{5}$

49. $\dfrac{2}{9} \div (-11)$

50. $\left(\dfrac{-1}{2}\right) \div \left(\dfrac{-1}{3}\right)$

Business application

51. *Elasticity* (*e*) measures the effect of price changes on consumer demand. The formula is

$$e = \dfrac{-P(\Delta Q)}{Q(\Delta P)}$$

where P is the price, Q is the demand, ΔP is the change in price, and ΔQ is the change in demand. Complete the following table.

e	P	Q	ΔP	ΔQ
?	10	200	$+1$	-40
?	40	800	-2	$+20$
?	200	60,000	$+10$	-300
?	100	20,000	$+5$	$-5,000$

Electricity application **52.** According to Ohm's law,

$$i = \frac{V}{R}$$

where i is the current in a simple circuit, V is the voltage, and R is the resistance. Complete the following table.

i	V	R
?	2000	200
?	-300	100
?	-80	20

Psychology application **53.** The z-score for an IQ is given by the formula

$$z = \frac{IQ - 100}{15}$$

Find the z-score for the following cases:
(a) $IQ = 130$
(b) $IQ = 85$
(c) $IQ = 55$
(d) $IQ = 100$

1.8   SUBSETS AND PROPERTIES OF THE REAL NUMBERS

Let us now look at the various types of numbers within the set of real numbers. These are summarized in the following table.

SET	ELEMENTS
N **(Natural numbers)**	$\{1, 2, 3, 4, 5, \ldots\}$ These are also called *counting numbers*.
I **(Integers)**	$\{\ldots, -3, -2, -1, 0, 1, 2, 3, \ldots\}$ These include the natural numbers, zero, and the negatives of the natural numbers.
Q **(Rational numbers)**	$\left\{\text{all quotients } \dfrac{a}{b},\ a \text{ and } b \text{ integers } (b \neq 0)\right\}$ This set includes all fractions, such as $\frac{1}{2}, (\frac{-5}{7}), 6 = \frac{6}{1}$. It also includes terminating decimals, such as $1.57 = \frac{157}{100}$. It also includes repeating decimals, such as $0.3333\ldots = \frac{1}{3}$.
H **(Irrational numbers)**	$\{\text{all nonterminating, nonrepeating decimals}\}$ These are the decimals, such as $2.483591\ldots$, that do not terminate or repeat. As examples, we have $\pi, \sqrt{2}, \sqrt{3}, \sqrt{5}$. (The square roots of numbers that are not perfect squares are irrational—more about this in Chapter 7.)
R **(Real numbers)**	$\{\text{all decimals}\} = \{\text{all points on number line}\} = \{\text{all rational or irrational numbers}\}$

The illustration shows the relation between these sets.

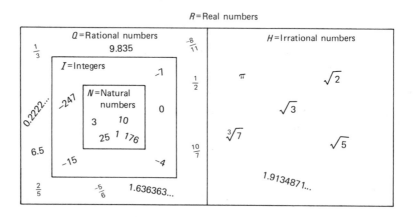

EXAMPLE 41   For each number, we list each of the special sets in which the number is a member.

	NUMBER	SETS CONTAINING NUMBER
(a)	$\dfrac{1}{2}$	Q, R
(b)	-7	I, Q, R
(c)	π	H, R
(d)	12	N, I, Q, R
(e)	$\dfrac{-10}{3}$	Q, R
(f)	$\sqrt{5}$	H, R
(g)	1.9	Q, R
(h)	$1.7777\ldots$	Q, R
(i)	$1.083623\ldots$	H, R
(j)	-18	I, Q, R

Notice that all the numbers listed are contained in R.

Throughout this chapter we have discussed some of the properties of the real numbers. Now, we state (or restate) the properties of the real numbers that we use later in the text. We give these properties names so that we can refer to them later. Throughout, a, b, and c are real numbers.

PROPERTY 6   (*Commutative Property*)   This property tells us that order does not affect an addition or a multiplication problem. In symbols,

Addition	$a + b = b + a$
Multiplication	$a \cdot b = b \cdot a$

PROPERTY 7   (*Associative Property*)   This property tells us that the grouping does not affect an addition or a multiplication problem. In symbols,

Addition	$a + (b + c) = (a + b) + c$
Multiplication	$a \cdot (b \cdot c) = (a \cdot b) \cdot c$

PROPERTY 8 (*Identity Property*)   We have already seen how 0 affects addition and 1 affects multiplication.

$$\text{Addition} \qquad 0 + a = a + 0 = a$$

$$\text{Multiplication} \qquad 1 \cdot a = a \cdot 1 = a$$

In words, adding 0 or multiplying by 1 does not change the value of any real number.

PROPERTY 9 (*Inverse Property*)   This property tells us that all real numbers have opposites, and all real numbers (except 0) have reciprocals.

$$\text{Addition} \qquad a + (-a) = 0$$

$$\text{Multiplication} \qquad a \cdot \left(\frac{1}{a}\right) = 1 \quad (a \neq 0)$$

PROPERTY 10 (*Distributive Property*)   This is the only property that involves both addition and multiplication at the same time.

$$a(b + c) = ab + ac$$

$$a(b - c) = ab - ac$$

We say *a distributes over b* and *c*. As examples,

$$4(\overset{\frown}{5 + 6}) = 4 \cdot 5 + 4 \cdot 6 = 20 + 24 = 44$$

$$3(\overset{\frown}{x + y}) = 3x + 3y$$

$$2(\overset{\frown}{a - 7}) = 2a - 14$$

EXAMPLE 42   The following are examples of the real-number properties stated above.

STATEMENT	PROPERTY
(a) $6 + 5 = 5 + 6$	Commutative property
(b) $7 \cdot 1 = 7$	Identity property
(c) $6(x + 3) = 6x + 18$	Distributive property

STATEMENT	PROPERTY
(d) $0 + 5 = 5$	Identity property
(e) $\dfrac{2}{3} \cdot \dfrac{3}{2} = 1$	Inverse property
(f) $6 \cdot (5 \cdot 2) = (6 \cdot 5) \cdot 2$	Associative property
(g) $-4 + 4 = 0$	Inverse property

PROBLEM SET 1.8

For each of the following numbers, list all the sets (N, I, Q, H, or R) to which it belongs.

See Example 41

1. $\dfrac{2}{5}$ **2.** -7 **3.** $\sqrt{5}$

4. 10 **5.** $\dfrac{-3}{8}$ **6.** 1.36

7. π **8.** 4 **9.** 0

10. -6 **11.** $-0.4444\ldots$ **12.** $2\tfrac{1}{2}$

State whether the following statements are true or false.

See Example 41

13. $\sqrt{2}$ is a rational number.

14. 5 is a rational number.

15. Every real number is a rational number.

16. Every natural number is an integer.

17. π is a real number.

18. -6 is a natural number.

19. Some rational numbers are integers.

20. Some rational numbers are irrational.

For each statement, give the real-number property that makes it true.

See Example 42

21. $6 + 0 = 6$ ~ Identity   **22.** $5(xy) = (5x)y$ Associative

23. $3 + 5 = 5 + 3$ Cummutative   **24.** $\dfrac{4}{5} \cdot \dfrac{5}{4} = 1$ inverse

25. $9(a - 2) = 9a - 18$ *distributive*   26. $8 + (-8) = 0$ *inverse*

27. $7 \cdot x = x \cdot 7$ *cummutative*   28. $1 \cdot u = u$ *identity*

29. $3(t + u) = 3t + 3u$ *distributive*   30. $(3 + r) + s = 3 + (r + s)$ *associative*

CHAPTER 1 SUMMARY

Important Words and Phrases

absolute value (1.3)
addition of real numbers (1.4)
additive inverse (1.3)
associative property (1.8)
brackets (1.1)
commutative property (1.8)
coordinate (1.3)
cubed (1.1)
difference (1.1)
distributive property (1.8)
division of real numbers (1.7)
equal sign (1.1)
equation (1.1)
evaluate (1.2)
exponent (1.1)
fraction bar (1.1)
grouping symbols (1.1)
identity property (1.8)
inequality (1.1)
inequality symbol (1.1)
integer (1.8)
inverse property (1.8)
irrational number (1.8)
multiplication of real numbers (1.6)

natural number (1.8)
negative real number (1.3)
negative sign (1.3)
number line (1.3)
operations (1.1)
opposites (1.3)
origin (1.3)
parentheses (1.1)
PEMDAS (1.1)
positive real numbers (1.3)
positive sign (1.3)
product (1.1)
quotient (1.1)
rational number (1.8)
real number (1.3)
reciprocal (1.7)
replacement set (1.2)
solution (1.2)
squared (1.1)
substitution property (1.2)
subtraction of real numbers (1.5)
sum (1.1)
variable (1.2)
zero (1.7)

Important Properties

Substitution Property. If $a = b$, then a can replace b, and vice versa, in any expression without changing its value. Also, a can replace b, and vice versa, in any statement without changing its truth or falsity.

For any real numbers a, b, and c:

$$a + b = b + a \qquad\qquad a \cdot b = b \cdot a$$

$$a + (b + c) = (a + b) + c \qquad a \cdot (b \cdot c) = (a \cdot b) \cdot c$$

$$0 + a = a + 0 = a \qquad\qquad 1 \cdot a = a \cdot 1 = a$$

$$a + (-a) = 0 \qquad\qquad a \cdot \frac{1}{a} = 1 \quad (a \neq 0)$$

$$a(b + c) = a \cdot b + a \cdot c$$

$$a(b - c) = a \cdot b - a \cdot c$$

$$-(-a) = a$$

$$0 \cdot a = 0$$

$$\frac{a}{b} = a\left(\frac{1}{b}\right)$$

We can *never* divide by zero: The fractions $\dfrac{a}{0}$ and $\dfrac{0}{0}$ are not defined.

Important Procedures

To add two real numbers:
1. With the *same* sign,
   (a) Add the absolute values.
   (b) Give the answer the common sign.
2. With *opposite* signs,
   (a) Subtract the absolute values (larger − smaller).
   (b) Give the answer the sign of the number with the larger absolute value.

To subtract two real numbers, add the opposite of the second number.

To multiply (divide) two real numbers:
1. With the *same* sign,
   (a) Multiply (divide) the absolute values.
   (b) The answer is positive.
2. With *opposite* signs,
   (a) Multiply (divide) the absolute values.
   (b) The answer is negative.

CHAPTER 1 REVIEW EXERCISES

State whether the following statements are true or false.

1. $2 \neq 7 - 4$ **2.** $-3 \geq 2$

3. 3 is a rational number

Translate the following phrases or sentences into algebraic symbols.

4. x is less than or equal to 10

5. the product of 5 and t

6. a is the sum of b and 8

What members of the replacement set make the given equation or inequality true?

7. $3x + 1 = 13$ $\{3, 4, 5\}$ **8.** $2a - 3 \leq 12$ $\{6, 7, 8\}$

Complete the following table.

	Number	Opposite	Reciprocal	Absolute value
9.	3			
10.	-7			
11.	$\dfrac{-2}{5}$			

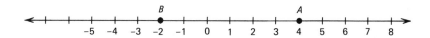

12. Give the coordinate of point A.

13. Give the coordinate of point B.

Simplify the following expressions as much as possible.

14. 5^2 **15.** 2^3

16. $8 - 2 \cdot 3$ **17.** $3(1 + 2)^2$

18. $5 + (-7)$ **19.** $-4 + (-8) + 9$

20. $-8 - (-9)$ **21.** $-7 - 2 - (-5) + (2 - 8)$

22. $(-5)(7)$ **23.** $(-2)(-6)$

24. $-2(1 - 7) - 4(-3 + 5)$ **25.** $\dfrac{-30}{-10}$

26. $\dfrac{-40}{8}$ **27.** $\dfrac{0}{0}$

28. $\dfrac{2(-5 - 3(2))}{1 - 9}$ **29.** $\left(\dfrac{-2}{3}\right) \div \left(\dfrac{-5}{7}\right)$

Evaluate the following expressions

30. $x + 4$ for $x = -8$ **31.** $t - 4$ for $t = -7$

32. $-6u$ for $u = -2$ **33.** $\dfrac{r}{-2}$ for $r = 16$

34. $(m + 5)(n - 4)$ for $m = -8$ and $n = 3$

35. $\dfrac{a - 2}{b + 3}$ for $a = -10$ and $b = -6$

What real-number properties make the following statements true?

36. $7(x - 3) = 7x - 21$ **37.** $0 + 6 = 6$

38. $\dfrac{-2}{5} \cdot \dfrac{-5}{2} = 1$

2

LINEAR EQUATIONS
AND INEQUALITIES

2.1   SIMPLIFYING EXPRESSIONS

In this chapter we solve equations and inequalities. We use the real-number properties discussed in Chapter 1 to help us solve the equations and inequalities.

The first step in solving any equation or inequality is to simplify it as much as possible. Let us begin with an example reviewing the distributive law.

EXAMPLE 1   Simplify $3(12 - x + 2y)$.

SOLUTION   We use the distributive law.

Distributive law

$3(12 - x + 2y) = 36 - 3x + 6y$

In any expression, a number or a product of a number and a variable is called a **term.** We simplify an expression by **combining like terms,** or terms with the same variable. For example, $5x$, $8x$, and $-3x$ are like terms.

EXAMPLE 2   Simplify the expression $2x + 5 + 8$.

SOLUTION   We combine the like terms, 5 and 8.

Combine terms

$2x + 5 + 8 = 2x + 13$

EXAMPLE 3   Simplify the expression $2x + 5x + 6$.

SOLUTION   We use the distributive law to combine the like terms, $2x$ and $5x$.

Distributive law

$2x + 5x + 6 = (2 + 5)x + 6$

Simplify

$= 7x + 6$

Often, we must group the like terms together first. For this, we use the commutative and associative laws of addition. (Recall that the commutative law allows us to reorder, and that the asssciative law allows us to regroup.)

EXAMPLE 4   Simplify the expression $T = 8x - 3 - 2x + 4 + 5x + 1$.

SOLUTION   We first rewrite all the subtractions as additions of opposite (change the signs and add). Then we regroup and simplify. We use the associative and commutative laws to regroup.

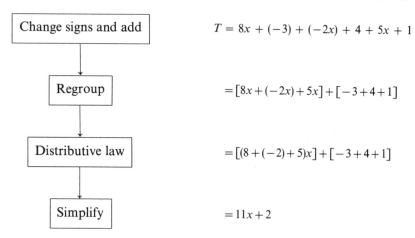

Notice that we group all the x-terms together and all the single numbers together.

EXAMPLE 5   Simplify the expression $T = -4 - 3x + 6x + 2 + 4x - 10 - 5x$.

SOLUTION   We first rewrite all the subtractions as additions of opposites; then we regroup and simplify.

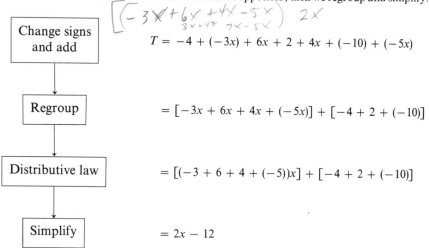

EXAMPLE 6   Simplify the expression $4(a - 7) + 3(4 - 2a) + 5$.

SOLUTION   We first use the distributive law to remove the parentheses; then we combine the like terms.

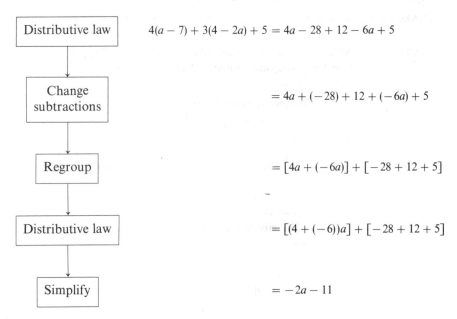

$4(a - 7) + 3(4 - 2a) + 5 = 4a - 28 + 12 - 6a + 5$

Distributive law

Change subtractions

$= 4a + (-28) + 12 + (-6a) + 5$

Regroup

$= [4a + (-6a)] + [-28 + 12 + 5]$

Distributive law

$= [(4 + (-6))a] + [-28 + 12 + 5]$

Simplify

$= -2a - 11$

PROBLEM SET 2.1

Simplify each of the following expressions as much as possible.

See Example 1

1. $4(x + 5)$ **2.** $6(a - 6)$

3. $5(2 - x)$ **4.** $-8(3 + 5x)$

5. $-3(4 - 2x + y)$ **6.** $7(8 + 3x - a)$

See Example 2

7. $7a + 5 + 3$ **8.** $8x - 4 + 6$

9. $6r + 3 - 2 - 8$ **10.** $-4x - 9 + 10 - 3$

11. $4p + 7 - 10 + 4$ **12.** $6y - 3 - 4 - 8$

See Example 3

13. $6x + 3x - 2$ **14.** $7x - 8x + 2$

15. $10a - 3a - 12$ **16.** $8u - 11u + 3$

17. $g - 3g + 8g - 5$ **18.** $6x - 7x - 9x + 7$

See Examples 4 and 5

19. $7x + 3 + 2x + 7$ **20.** $4x + 2 + 3x + 7$

21. $8a - 7 - 2a - 3$ **22.** $4x - 2 - 9x + 3$

23. $4u - 10 - u + 3 - 5u$ **24.** $7r + 11 - 5r + 17 - 4r$

25. $3 - x - 7 + 4x - 12$ **26.** $7 - x - 3x - 4 + 8x - 3$

27. $4k - 6 + 7k - 8 - 3k + 2$ **28.** $7 - x - 3x - 8 + 9x - 3$

29. $8m - 9 - 10m + m - 3 - m$ **30.** $8t - 2t + 6 - 9t - 14 + t$

See   **31.** $2(x + 5) - 3$ **32.** $8(a - 2) + 9$

Example 6

33. $4(2x - 3) + 6x$ **34.** $6(2 - 3x) - x$

35. $2(u - 5) + 3(2u + 1)$ **36.** $4(6 + t) - 3(t - 5)$

37. $3(x - 2) - 4(5 - 2x)$ **38.** $-2(x - 5) + 5(3 + 2x)$

39. $6(m - 8) + 2(3 - m) - 7m + 2$

40. $3(k - 2) - 8(1 - k) + 4k - 6$

2.2 ADDITION PROPERTY OF EQUALITY

The equation $2 + 3 = 5$ is true.

The equation $2 + 5 = 6$ is false.

How about the equation $2 + x = 7$? True or false? The answer depends on x: if $x = 5$, it is true; if $x \neq 5$, it is false. The values that make such equations true are called the **solutions** (or if you use sets, the **solution set**). This is what we spend the rest of the chapter doing—finding the solutions for equations and inequalities.

Two or more equations are called **equivalent** if they have the same solution.

EXAMPLE 7   The following equations are all equivalent since they all have the same solution, 5. In other words, if $x = 5$, then each of the equations is true.

(a) $x = 5$ **(b)** $x + 3 = 8$

(c) $2x = 10$ **(d)** $\dfrac{x}{5} = 1$

(e) $3x + 1 = 16$ **(f)** $16 - x = x + 6$

Of these equations, notice that $x = 5$ is the simplest, since the solution is obvious. In general, we always look for the simplest equation, which has the solution isolated as

$$x = \text{number}$$

To help us isolate the unknown, we use the **addition property of equality.**

PROPERTY 1 (*Addition Property of Equality*)   Let A, B, and C be algebraic expressions. If $A = B$, then $A + C = B + C$.

In words, we can *add the same number to both sides of an equation, and the results are still equal.* Since subtracting and adding the opposite are the same, we can also *subtract the same number from both sides of an equation, and the results will still be equal.*

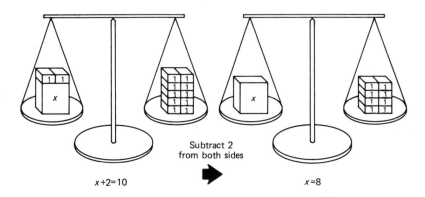

$x+2=10$ Subtract 2 from both sides $x=8$

The illustration shows the equation $x + 2 = 10$ on the left. As a scale, it balances. If we remove (subtract) two units from both sides of the scale, it will still balance. But now we see that the unknown block, x, is 8.

EXAMPLE 8   Solve $x - 4 = 11$ for x.

SOLUTION   Here we wish to isolate x. Since 4 is subtracted from x, we add 4 to both sides. This will produce $-4 + 4$ on the left, which will cancel, leaving x alone.

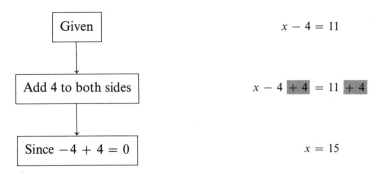

Given	$x - 4 = 11$
Add 4 to both sides	$x - 4 \; \boxed{+\,4} = 11 \; \boxed{+\,4}$
Since $-4 + 4 = 0$	$x = 15$

The solution set is {15}, or we can say that the solution is 15. We can check this: $15 - 4 = 11$.

EXAMPLE 9   Solve $a + 6 = 20$.

SOLUTION   Here, a is not alone. We can remove the 6 that is added to a by subtracting 6 from both sides.

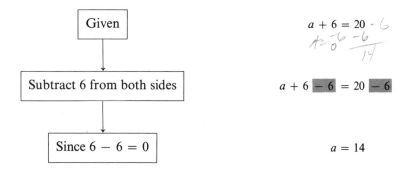

The solution is 14. We can check this: $14 + 6 = 20$.

EXAMPLE 10   Solve $5x - 4x - 7 = 12$.

SOLUTION   We first simplify the left-side expression; then we solve.

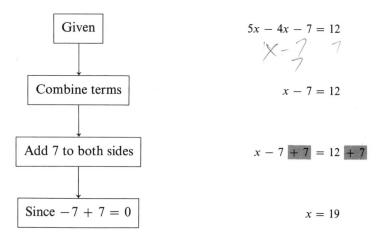

The solution is 19.

EXAMPLE 11   Solve $8t - 3 = 7t + 5$.

SOLUTION   We use the addition property of equality (Property 1) to move all the *t*-terms to the left side of the equation, and all the number terms to the right side.

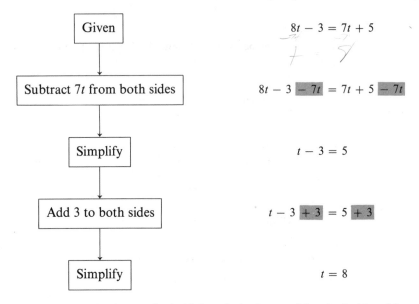

The solution is $t = 8$. We can check this by substituting $t = 8$ into both sides of the original equation.

$$Left\ side\ = 8(8) - 3 = 64 - 3 = 61$$

$$Right\ side = 7(8) + 5 = 56 + 5 = 61 \qquad Checks.$$

PROBLEM SET 2.2

For each set of equations, determine whether or not they are equivalent (same solution set).

See
Example 7

1. $x = 2$
        $4x = 8$
      $3x + 2 = 5x - 2$
      $10 - x = 6 + x$

2. $x = 3$
        $3x = 9$
      $4x - 1 = 2x + 3$
      $12 - x = 7 + x$

3. $x = -4$
       $-3x = 12$
      $8 - 2x = 4 + 6x$
      $3 - x = -2x - x$

4. $a = -5$
        $4a = -20$
      $10 - a = -3a$
      $3a + 8 = a - 2$

Solve the following equations by simplifying and by using the addition property of equality (Property 1).

See Example 8

5. $x - 3 = 12$ **6.** $a - 6 = 10$

7. $t - 8 = 2$ **8.** $x - 4 = -8$

9. $y - 2 = -1$ **10.** $u - 10 = 21$

See Example 9

11. $x + 7 = 11$ **12.** $a + 2 = 5$

13. $r + 8 = 5$ **14.** $y + 7 = 1$

15. $x + 12 = -2$ **16.** $t + 4 = -6$

See Example 10

17. $3x - 2x - 7 = 2$ **18.** $8x - 7x + 2 = 5$

19. $10t + 3 - 9t = 6$ **20.** $4y - 5 - 3y = 7$

21. $16a - 2 - 15a = -4$ **22.** $2a + 5 - a = -6$

See Example 11

23. $5x = 4x + 6$ **24.** $9a = 8 + 8a$

25. $10t - 1 = 9t + 8$ **26.** $2y - 5 = y - 8$

27. $6u + 5 = 5u - 3$ **28.** $11x + 7 = 10x - 18$

Business application

29. The formula for profit (P) is given by

$$P = R - C$$

where R is the revenue and C is the cost. Find R if $P = \$20,000$ and $C = \$100,000$.

Science application

30. The energy (E) in a moving object is given by

$$E = K + P$$

where K is the kinetic energy and P is the potential energy. Find K if $E = 150$ joules and $P = 95$ joules.

2.3   MULTIPLICATION PROPERTY OF EQUALITY

Just as we can add the same number to both sides of an equation, we can *multiply both sides of an equation by the same nonzero number, and the results will still be equal.*

> **PROPERTY 2** (*Multiplication Property of Equality*)   Let A, B, and C be algebraic expressions ($C \neq 0$). If $A = B$, then $AC = BC$.

EXAMPLE 12   Solve $6x = 18$.

SOLUTION   To get the x alone, we multiply both sides by $\frac{1}{6}$ to remove the $6\left[\text{ since }\frac{1}{6}(6) = 1\right]$.

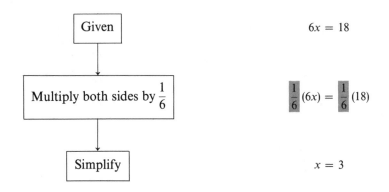

In Example 12 it might be reasonable to have divided both sides by 6. Since multiplying both sides by $\frac{1}{C}$ is the same as dividing by C, we can say that we can *divide both sides of an equation by the same nonzero number, and the results will also be equal.*

EXAMPLE 13   Solve $-4a = 100$.

SOLUTION   Instead of multiplying both sides by $\frac{-1}{4}$, we divide both sides by -4. This gives the same result.

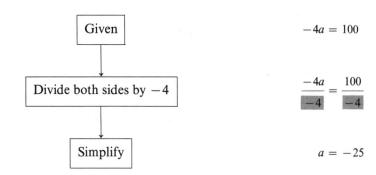

EXAMPLE 14   Solve $\dfrac{k}{5} = 7$.

SOLUTION   Here we multiply both sides by 5 to get k alone.

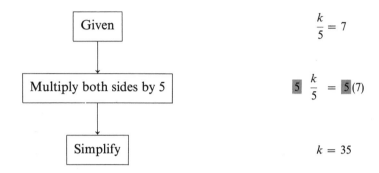

EXAMPLE 15   Solve $-x = 12$.

$$-x \quad -x$$
$$x = -12$$

SOLUTION   This almost looks like an answer, but it is not. We want x, *not* $-x$. Think of $-x$ as $-1(x)$, so that multiplying by -1 will produce x.

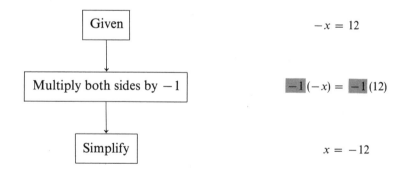

EXAMPLE 16   Solve $\dfrac{3}{5}m = 36$.

SOLUTION   We want to get m alone. Since $\dfrac{3}{5} \cdot \dfrac{5}{3} = 1$, we multiply both sides by $\dfrac{5}{3}$.

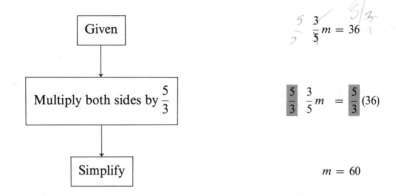

Given	$\frac{3}{5} m = 36$
Multiply both sides by $\frac{5}{3}$	$\frac{5}{3} \frac{3}{5} m = \frac{5}{3} (36)$
Simplify	$m = 60$

EXAMPLE 17   Solve $8x + 2x - 3 = 7$.

SOLUTION   We first simplify the expression on the left; then we isolate x and solve.

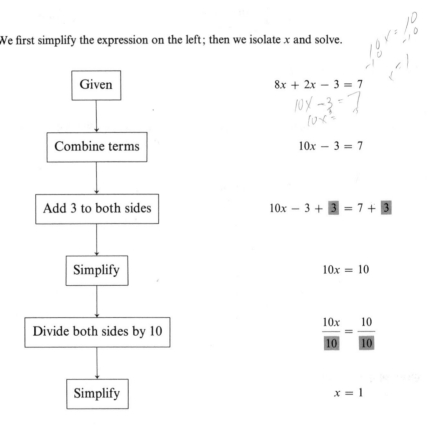

Given	$8x + 2x - 3 = 7$
Combine terms	$10x - 3 = 7$
Add 3 to both sides	$10x - 3 + 3 = 7 + 3$
Simplify	$10x = 10$
Divide both sides by 10	$\frac{10x}{10} = \frac{10}{10}$
Simplify	$x = 1$

EXAMPLE 18   Solve $8y + 1 = 5y - 5$.

SOLUTION We use the addition and subtraction properties to get all the y-terms together on the left and all the numbers on the right.

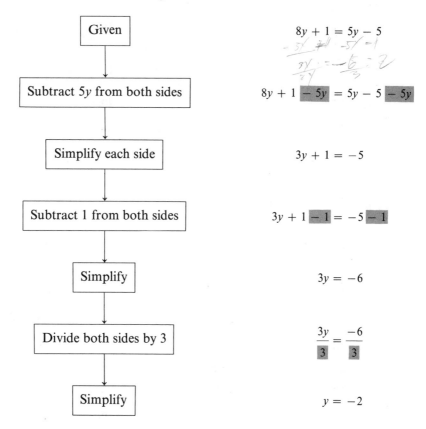

| Given | $8y + 1 = 5y - 5$ |

| Subtract $5y$ from both sides | $8y + 1 \boxed{-5y} = 5y - 5 \boxed{-5y}$ |

| Simplify each side | $3y + 1 = -5$ |

| Subtract 1 from both sides | $3y + 1 \boxed{-1} = -5 \boxed{-1}$ |

| Simplify | $3y = -6$ |

| Divide both sides by 3 | $\dfrac{3y}{\boxed{3}} = \dfrac{-6}{\boxed{3}}$ |

| Simplify | $y = -2$ |

PROBLEM SET 2.3

Solve the following equations.

See Examples 12 and 13

1. $2x = 10$ **2.** $3a = 15$

3. $5x = 20$ **4.** $7t = 21$

5. $-6a = 42$ **6.** $-10x = -80$

7. $3k = -24$ **8.** $-4m = 16$

9. $-6x = -6$ **10.** $3y = -12$

See Example 14

11. $\dfrac{x}{4} = 3$ **12.** $\dfrac{a}{6} = 5$

13. $\dfrac{r}{5} = 12$

14. $\dfrac{m}{10} = -2$

15. $\dfrac{x}{-2} = 8$

16. $\dfrac{z}{4} = -5$

17. $\dfrac{k}{-3} = -2$

18. $\dfrac{x}{-6} = -1$

See Example 15

19. $-x = 4$

20. $-a = 2$

21. $-t = -7$

22. $-x = -8$

23. $-m = \dfrac{3}{5}$

24. $-k = \dfrac{-4}{3}$

See Example 16

25. $\dfrac{2}{3}x = 10$

26. $\dfrac{3}{4}x = 15$

27. $\dfrac{2}{9}a = 8$

28. $\dfrac{4}{3}k = 16$

29. $\dfrac{-2}{7}t = 6$

30. $\dfrac{-5}{4}x = -20$

See Example 17

31. $4x + 2x - 1 = 5$

32. $3a + 2a + 4 = 14$

33. $2x + 5 + 6x = -3$

34. $10k - 4 - 3k = 10$

35. $3m - 5m + 2 = 8$

36. $z - 6z - 3 = 2$

See Example 18

37. $8x - 1 = 3x + 9$

38. $4a - 7 = a + 8$

39. $4t - 6 = t - 3$

40. $3z - 8 = 5z - 2$

41. $6x - 5 = 4x + 3$

42. $10k + 3 = 7k - 9$

43. $4x + 7 = 9x + 2$

44. $6a - 4 = 9a + 8$

Health application

45. Young's formula for child's dosage C of medicine is given by

$$C = \left(\dfrac{A}{A + 12}\right)D$$

where A is the child's age and D is the adult dosage. Find D if $C = 50$ and $A = 6$ years.

Consumer application

46. The cost C to operate a certain automobile for one year is given by

$$C = 1200 + 0.08m$$

where m is the number of miles driven in 1 year. Find m if $C = \$2000$.

Business applications **47.** The average cost A to produce Q items is given by

$$A = \frac{C}{Q}$$

where C is the total cost. Find the total cost C if $A = \$10$ and $Q = 500$ items.

48. The relation between the demand Q for an item and its price P is given by

$$Q + 1000P = 6000$$

Find P if $Q = 2000$ items.

Science application **49.** The relation between Fahrenheit (F) temperature and Celsius (C) is given by

$$F = \tfrac{9}{5} C + 32$$

Find C if $F = 68°$.

2.4   SOLVING LINEAR EQUATIONS

In this section we use all the methods shown in the first three sections to solve the **general linear equation.** This is an equation with only numbers and an unknown variable, but no products such as $x \cdot x$ or $x \cdot y$. Here is a general procedure that we use.

To solve linear equations:

1. Use the distributive law to remove all parentheses and brackets.
2. Simplify the left and right sides of the equation by combining like terms.
3. Move all terms with the unknown to one side of the equation.
4. Move all the number terms to the other side of the equation.
5. Multiply or divide to get the unknown by itself.
6. Check by substituting the solution into the original equation.

(If a step is not needed, skip it. Be careful with negative numbers.)

EXAMPLE 19   Solve $3x + 9x + 5 = 7x + 20$.

SOLUTION   We first simplify the left side of the equation. Then we get the x-terms together, the number terms together, and solve.

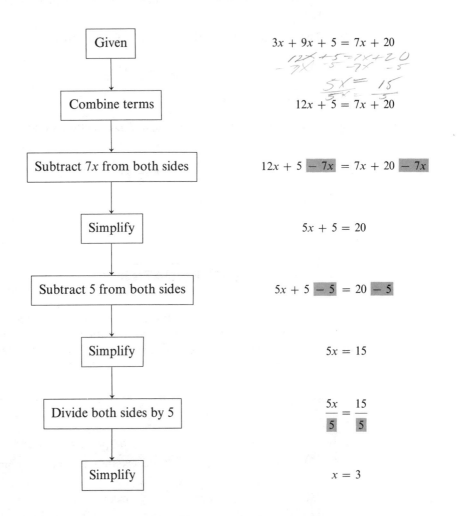

Given	$3x + 9x + 5 = 7x + 20$
Combine terms	$12x + 5 = 7x + 20$
Subtract $7x$ from both sides	$12x + 5 \; -7x = 7x + 20 \; -7x$
Simplify	$5x + 5 = 20$
Subtract 5 from both sides	$5x + 5 \; -5 = 20 \; -5$
Simplify	$5x = 15$
Divide both sides by 5	$\dfrac{5x}{5} = \dfrac{15}{5}$
Simplify	$x = 3$

We can check this by substituting $x = 3$ into the left and right sides of the original equation.

$$Left\ side\ = 3(3) + 9(3) + 5 = 9 + 27 + 5 = 41$$
$$Right\ side = 7(3) + 20 = 21 + 20 = 41 \quad Checks.$$

EXAMPLE 20   Solve $2(x + 5) = 5x + 16$.

SOLUTION   We first remove the parentheses using the distributive law. Then we get the x-terms on the left and the number terms on the right.

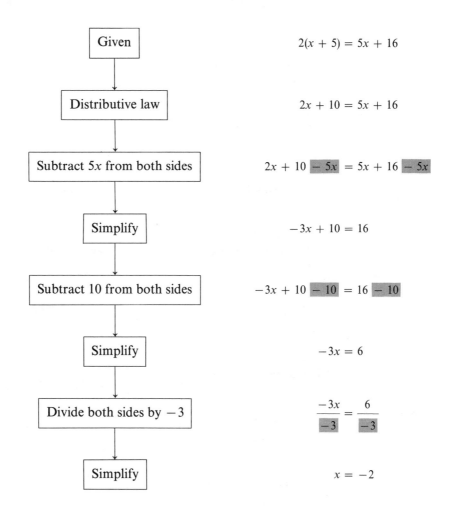

We can check this by substituting.

$$\textit{Left side} \ = 2(-2 + 5) = 2(3) = 6$$

$$\textit{Right side} = 5(-2) + 16 = -10 + 16 = 6 \qquad \textit{Checks.}$$

EXAMPLE 21   Solve $4(2x - 3) = 6(12 - x)$.

SOLUTION   Again, we begin by using the distributive law to clear away the parentheses.

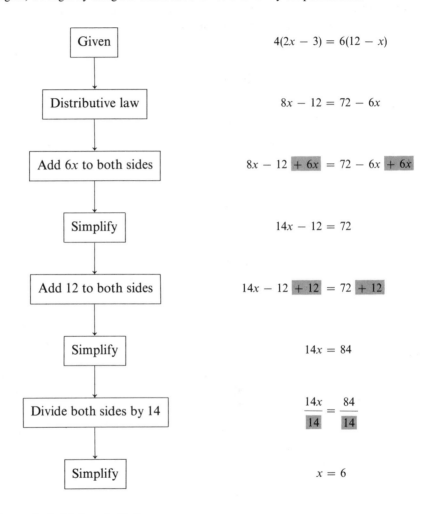

Given	$4(2x - 3) = 6(12 - x)$
Distributive law	$8x - 12 = 72 - 6x$
Add $6x$ to both sides	$8x - 12 \boxed{+ 6x} = 72 - 6x \boxed{+ 6x}$
Simplify	$14x - 12 = 72$
Add 12 to both sides	$14x - 12 \boxed{+ 12} = 72 \boxed{+ 12}$
Simplify	$14x = 84$
Divide both sides by 14	$\dfrac{14x}{\boxed{14}} = \dfrac{84}{\boxed{14}}$
Simplify	$x = 6$

Let us check this by substituting:

$$\textit{Left side}\ \ = 4[2(6) - 3] = 4[12 - 3] = 4(9) = 36$$
$$\textit{Right side} = 6(12 - 6) = 6(6) = 36 \quad \textit{Checks.}$$

EXAMPLE 22   Solve $5(3 - a) - 4 = 3(3a - 1) - 21a$.

SOLUTION   We use the distributive law, combine like terms, collect the a-terms on the left and the numbers on the right and solve.

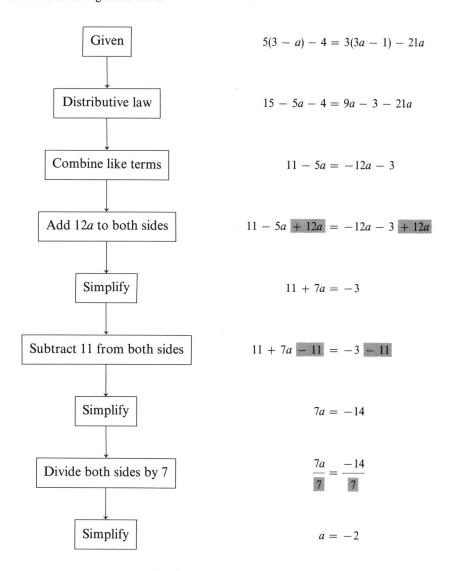

| Given | $5(3 - a) - 4 = 3(3a - 1) - 21a$ |

| Distributive law | $15 - 5a - 4 = 9a - 3 - 21a$ |

| Combine like terms | $11 - 5a = -12a - 3$ |

| Add 12a to both sides | $11 - 5a \ +12a = -12a - 3 \ +12a$ |

| Simplify | $11 + 7a = -3$ |

| Subtract 11 from both sides | $11 + 7a \ -11 = -3 \ -11$ |

| Simplify | $7a = -14$ |

| Divide both sides by 7 | $\dfrac{7a}{7} = \dfrac{-14}{7}$ |

| Simplify | $a = -2$ |

We check this by substituting $a = -2$ into the original equation.

$$Left\ side\ = 5[3 - (-2)] - 4 = 5(5) - 4 = 25 - 4 = 21$$

$$Right\ side = 3[3(-2) - 1] - 21(-2) = 3(-6 - 1) + 42$$

$$= 3(-7) + 42 = -21 + 42 = 21 \quad Checks.$$

YES	NO
$5(3 - a) - 4 = 15 - 5a - 4$	~~$5(3 - a) - 4 = 15 - 5a - 20$~~

We distribute only over the terms *within* the parentheses.

PROBLEM SET 2.4

Solve the following equations and check the solutions.

See Example 19

1. $2x - 3 = 5x - 18$ **2.** $7x + 5 = 10x + 2$

3. $6a + 10 = 2a - 10$ **4.** $8k - 25 = 5k + 5$

5. $7m + 11 = 3m - 1$ **6.** $4x + 20 = 7x - 1$

7. $6x - 9 = 17x + 2$ **8.** $3a + 4 = 5a - 20$

See Example 20

9. $2(x + 5) = 7x - 5$ **10.** $6(x - 3) = 4x + 2$

11. $6(2a - 1) = 10a + 4$ **12.** $4(3k - 5) = 7k - 35$

13. $3(5x - 2) = 11x + 34$ **14.** $10(3 - m) = -4m - 6$

15. $5(6 - y) = -12y + 2$ **16.** $4(6 - 2x) = 12x + 4$

See Example 21

17. $4(x + 2) = 7(x - 1)$ **18.** $3(x + 7) = 9(x + 1)$

19. $2(2x - 1) = 7(x - 2)$ **20.** $6(4 - a) = 9(1 - a)$

21. $5(6 - 3k) = 2(4 - 2k)$ **22.** $4(a - 7) = 2(a - 4)$

23. $3(5x - 2) = 4(2x + 2)$ **24.** $10(4m + 3) = 11(6m - 2)$

See Example 22

25. $4(a + 2) + 3a = 5(a - 3) + 31$

26. $5(x - 5) + 7x = 4(x + 2) - 9$

27. $6(2k - 1) - 5k = 4(5k + 2) + 51$

28. $8(3y - 2) + 7y = -4(5 - y) - 23$

29. $4(2x + 5) - 5(3x - 1) = 46$

30. $7(3x + 1) - 6(2x + 5) = 13$

31. $2(7a - 1) + 3(3a + 2) = 20(a + 1) - 1$

32. $5(3r - 2) - 7(r + 6) = 9(4 - r) + 14$

Business applications **33.** The relation between price P and demand Q for a certain item is given by

$$Q = 500(10 - P)$$

Find P if $Q = 2000$.

34. In Illinois, the income tax T that a family pays is given by

$$T = 0.025(I - 1000E)$$

where I is income and E is the number of exemptions. Find I if $T = \$300$ and $E = 3$.

Science applications **35.** According to the first law of thermodynamics, the heat (Q) added to a boiling process is given by

$$Q = U + p(V_V - V_L)$$

where U is the change in internal energy, p is the pressure, V_V is the volume of vapor, and V_L is the volume of liquid. Find V_L if $Q = 540$, $U = 500$, $p = 2.5$, and $V_V = 20$.

36. The relation between Fahrenheit (F) and Celsius (C) temperatures can be written

$$C = \tfrac{5}{9}(F - 32)$$

Find F if $C = 20°$.

2.5   WORD PROBLEMS

Many algebra problems start as word problems. These are English sentences that must be translated into algebraic symbols and equations. These problems do not have to be as scary as some students imagine them to be. In Section 1.1 we talked briefly about translating English into algebra. Here we present a table of translations that we use in these word problems. (This table is very similar to the translation tables of Section 1.1, but is more complete.)

English	Algebra
plus, more than, added to, increased by, sum of, total of	$+$
minus, less than, decreased by, subtracted from, difference of	$-$
times, of, product of, multiplied by	$\cdot$
divided by, per, ratio of, quotient of	$\div$
is, are, was, should be, represents	$=$
what, some number	x
consecutive integers	$x, x + 1, x + 2$
consecutive odd (or even) integers	$x, x + 2, x + 4$

To solve word problems:

1. Read (and reread) the problem carefully.
2. Identify the unknown (or unknowns) by a letter (or letters).
3. Translate the English information into an algebraic equation.
4. Solve the equation.
5. Check the solution.

EXAMPLE 23   One number is 3 more than another number. Their sum is 41. Find the numbers.

SOLUTION We have two numbers here. We can call one of them x. Since the other is 3 *more than* x, it is $x + 3$.

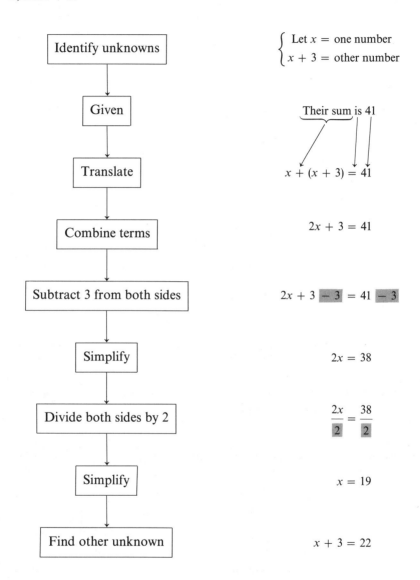

$$\begin{cases} \text{Let } x = \text{one number} \\ x + 3 = \text{other number} \end{cases}$$

Their sum is 41

$$x + (x + 3) = 41$$

$$2x + 3 = 41$$

$$2x + 3 - 3 = 41 - 3$$

$$2x = 38$$

$$\frac{2x}{2} = \frac{38}{2}$$

$$x = 19$$

$$x + 3 = 22$$

The numbers are 19 and 22. We check: 22 is 3 more than 19, and their sum is 41.

EXAMPLE 24 Fifty is 2% of what number?

SOLUTION   This is a percent problem of the type often seen in arithmetic. We translate this as we did in Example 23. Also, recall that $2\% = 0.02$.

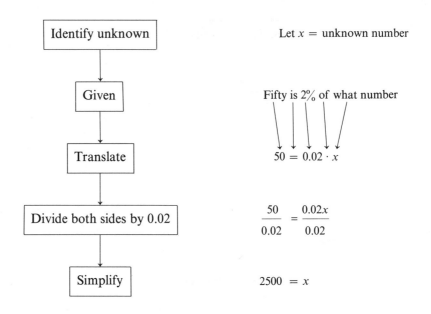

The solution is 2500. We can check this: 2% of 2500 is $(0.02)(2500) = 50$.

EXAMPLE 25   Larry and Nancy earn \$23,300 together. Larry earns \$900 more than Nancy. How much does each of them earn?

SOLUTION   If Nancy's salary is x, then Larry's is 900 more, or $x + 900$. The total is 23,300.

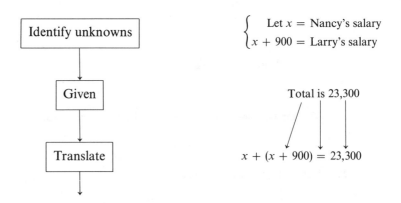

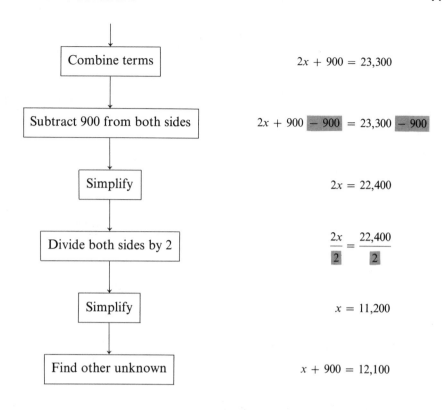

Combine terms $2x + 900 = 23{,}300$

Subtract 900 from both sides $2x + 900 - 900 = 23{,}300 - 900$

Simplify $2x = 22{,}400$

Divide both sides by 2 $\dfrac{2x}{2} = \dfrac{22{,}400}{2}$

Simplify $x = 11{,}200$

Find other unknown $x + 900 = 12{,}100$

Therefore, Nancy's salary is \$11,200 and Larry's is \$12,100. Together, this checks as a \$23,300 total.

EXAMPLE 26 The perimeter of a rectangle is 26. The length is one more than three times the width. Find the dimensions.

SOLUTION Let W be the width. (It is usually easier to identify the letter as the smaller number.) Recall that the perimeter of a rectangle is $P = 2L + 2W$.

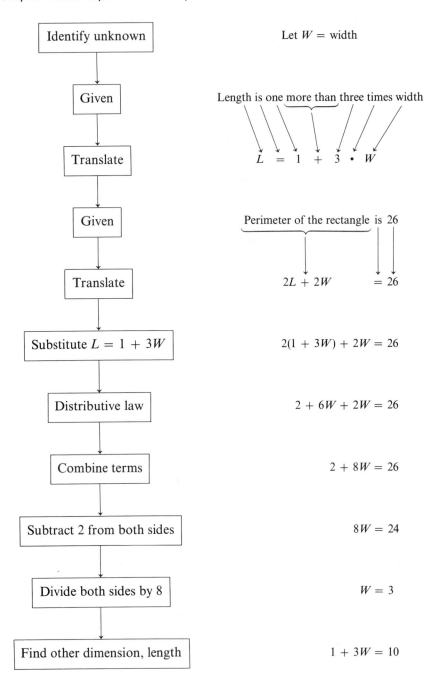

The dimensions are 3 and 10. The perimeter is $2(3) + 2(10) = 26$.

PROBLEM SET 2.5

Solve the following word problems.

*See
Example 23*

1. One number is 8 more than another number. If their sum is 42, find the numbers.

2. One number is 12 less than another number. If their sum is 90, find the numbers.

3. One number is 5 times another number. If their difference is 28, find the numbers.

4. One number is one-third of another number. If their sum is 48, find the numbers.

5. The sum of two consecutive integers is 47. Find the numbers.

6. The sum of two consecutive odd integers is 56. Find the numbers.

*See
Example 24*

7. What is 8% of 500?

8. What is $12\frac{1}{2}\%$ of 6400?

9. Eighty-one is what percent of 540?

10. Five hundred is what percent of 400?

11. Two hundred eighty-seven is 35% of what number?

12. Three hundred seventy-five is 3% of what number?

*See
Example 25*

13. In a certain condominium, there are twice as many two-bedroom units as there are three-bedroom units. There are a total of 114 units. How many of each are there?

14. In another condominium, there are the same number of one- and two-bedroom units, and 40 fewer three-bedroom units. If there are 230 total units, how many of each are there?

15. Bob has 1000 feet of recording tape. He cuts it into two pieces: one 150 feet longer than the other. How long is each piece?

16. The front-four of the Tigers' defense is made of four very heavy players. Jones is 10 pounds heavier than Smith, who is 5 pounds heavier than Wilson, who is 10 pounds heavier than Green. Together, they weigh 990 pounds. How much does each of them weigh?

*See
Example 26*

17. The perimeter of a square is 92. Find the length of a side.

18. The lengths of the sides of a triangle are three consecutive odd integers. The perimeter is 87. Find the length of each side.

19. The width of a rectangle is 6 less than the length. The perimeter is 56. Find the length and width.

20. The length of a rectangle is 2 less than five times the width. The perimeter is 56. Find the length and the width.

2.6   LINEAR INEQUALITIES

Solving **linear inequalities** is similar to solving linear equations, except that

1. Instead of $=$, we will be dealing with one of the four inequality symbols: $<$, $\leq$, $>$, or $\geq$.

2. Instead of a single-number solution, our solution set will be a half-line that we will graph.

3. The multiplication property is slightly trickier, as we will see.

First, we have the **addition property of inequalities.**

> *PROPERTY 3* (*Addition Property of Inequalities*)   Let A, B, and C be algebraic expressions. If $A < B$, then $A + C < B + C$ (similarly for $\leq$, $>$, and $\geq$).

In words, we can *add* (*or subtract*) *the same number to* (*or from*) *both sides of an inequality, and the inequality has the same direction.*

EXAMPLE 27   Solve $x + 5 < 12$.

SOLUTION   Just as with linear equations, we isolate x by subtracting 5 from both sides.

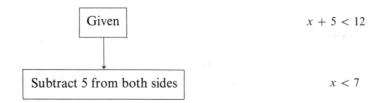

| Given | $x + 5 < 12$ |

| Subtract 5 from both sides | $x < 7$ |

We graph this solution as follows.

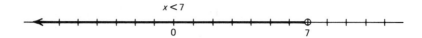

We use an open dot at $x = 7$ to indicate that it is *not* in the solution set.

EXAMPLE 28   Solve $a - 6 \geq 5$.

SOLUTION   We add 6 to both sides.

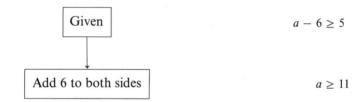

We graph this solution as follows.

Here we use a solid dot at $a = 11$ to indicate that it *is* in the solution set.

The multiplication property is not as simple. The rule depends on whether we are multiplying by a positive or negative number. Consider the following cases.

Inequality	Multiply by 4
$2 < 5$	$8 < 20$
$-2 < 3$	$-8 < 12$
$-3 < -1$	$-12 < -4$

Notice that multiplying by 4 kept the inequality the same.

INEQUALITY	MULTIPLY BY -4
$2 < 5$	$-8 > -20$
$-2 < 3$	$8 > -12$
$-3 < -1$	$12 > 4$

Notice that multiplying the same inequalities by -4 *reverses* the direction of the inequality. This suggests the **multiplication property of inequalities.**

> **_PROPERTY 4_** (*Multiplication Property of Inequalities*)   Let A, B, and C be algebraic expressions.
>
> **1.** For $C > 0$, if $A < B$, then $AC < BC$.
> **2.** For $C < 0$, if $A < B$, then $AC > BC$.
>
> (Similarly, this is true for $\leq$, $>$, and $\geq$.)

In words, *multiplying (or dividing) both sides of an inequality by a **positive** number does not change the inequality direction*; on the other hand, *multiplying (or dividing) by a **negative** number reverses the inequality direction.*

EXAMPLE 29   Solve $3x < 15$.

SOLUTION   Since $3 > 0$, the inequality will not change when we divide both sides by 3.

| Given | $3x < 15$ |

| Divide both sides by 3 | $x < 5$ |

The solution appears as follows.

EXAMPLE 30   Solve $-5a \le 20$.

SOLUTION   Since $-5 < 0$, the inequality will reverse when we divide both sides by -5.

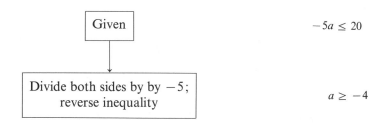

The solution can be graphed as follows.

EXAMPLE 31   Solve $\dfrac{-x}{6} > -1$.

SOLUTION   To isolate x we multiply both sides of the inequality by -6. This will reverse the inequality.

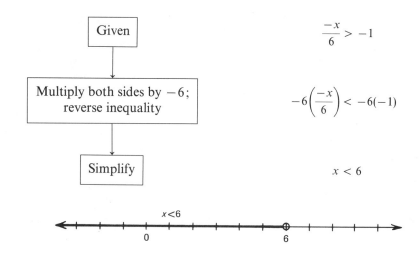

EXAMPLE 32   Solve $2x + 5 < 8x - 7$.

SOLUTION   Just as with linear equations, we collect all the x-terms on one side and all the numbers on the other side.

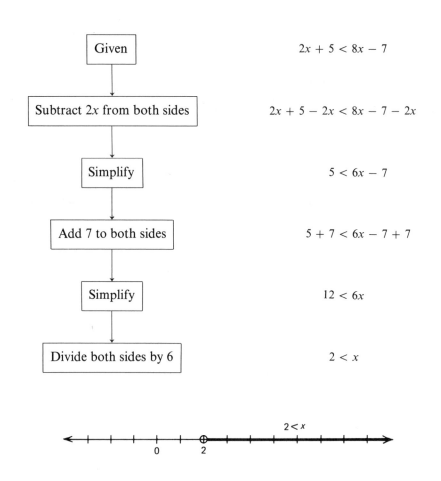

Given	$2x + 5 < 8x - 7$
Subtract $2x$ from both sides	$2x + 5 - 2x < 8x - 7 - 2x$
Simplify	$5 < 6x - 7$
Add 7 to both sides	$5 + 7 < 6x - 7 + 7$
Simplify	$12 < 6x$
Divide both sides by 6	$2 < x$

EXAMPLE 33   Solve $2x + 3(4 - x) \geq 2(x + 5) + 11$.

SOLUTION   Just as with equations, we first clear the parentheses; then we combine terms and solve.

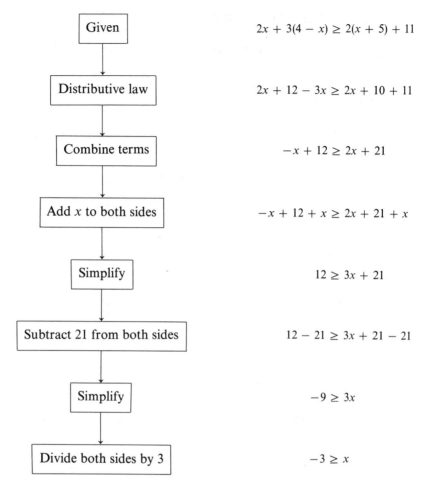

Given	$2x + 3(4 - x) \geq 2(x + 5) + 11$
Distributive law	$2x + 12 - 3x \geq 2x + 10 + 11$
Combine terms	$-x + 12 \geq 2x + 21$
Add x to both sides	$-x + 12 + x \geq 2x + 21 + x$
Simplify	$12 \geq 3x + 21$
Subtract 21 from both sides	$12 - 21 \geq 3x + 21 - 21$
Simplify	$-9 \geq 3x$
Divide both sides by 3	$-3 \geq x$

The graph of this solution is as follows.

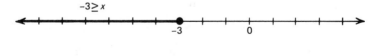

YES	NO
$2x < -10$	$2x < -10$
$x < -5$	$x > -5$

We only reverse the inequality when we multiply or divide *by* a negative number, not *into* a negative number.

PROBLEM SET 2.6

Solve the following inequalities and graph the solutions.

See
Examples
27 and 28

1. $x + 4 < 8$ **2.** $a + 7 > 4$

3. $k - 9 \geq 2$ **4.** $y + 2 \leq -6$

5. $m - 10 > -5$ **6.** $x - 1 < 18$

7. $x - 4 \leq 3$ **8.** $k + 6 \geq 10$

9. $a - 14 < 19$ **10.** $r + 5 > -4$

11. $y - 6 \geq -1$ **12.** $x - 10 \leq 21$

See
Examples
29 and 30

13. $3x > 30$ **14.** $5a < 35$

15. $-2k \leq 6$ **16.** $-4t \geq 24$

17. $6m < -12$ **18.** $10x > -30$

19. $-8y \geq -16$ **20.** $-2u \leq -40$

See
Example 31

21. $\dfrac{x}{2} > 3$ **22.** $\dfrac{a}{5} < 1$

23. $\dfrac{-x}{4} \leq 10$ **24.** $\dfrac{-y}{6} \geq 3$

25. $\dfrac{t}{4} < -5$ **26.** $\dfrac{m}{3} > -2$

27. $\dfrac{-f}{2} \geq -3$ **28.** $\dfrac{-x}{5} \leq -1$

29. $2x + 6 > 4x - 8$ **30.** $3x - 5 < x + 3$

See
Example 32

31. $7a - 2 \leq 5a - 6$ **32.** $4u - 11 \geq 7u - 20$

33. $5y + 13 < 7 - y$ **34.** $6x - 15 > 2x + 25$

35. $2x + 13 \geq -29 - 5x$ **36.** $3x + 45 \leq 5 - x$

See
Example 33

37. $2(x + 5) > 7(4 - x)$ **38.** $8(x - 3) < 4(x - 1)$

39. $5(2x - 1) \leq -3(3 - 2x)$ **40.** $2(5 - x) \geq 4(3x - 8)$

41. $3x + 5(x - 7) < 4(x - 1) - 27$

42. $5u - 3(8 - u) \geq -7(3 - 2u) + 15$

43. $8a - 6(2a + 1) > 4(5a + 3) + 222$

44. $5m - 2(4m - 3) \le 2(3 - 6m) - 18$

Business application **45.** A manufacturing firm has 320 labor-hours to spend producing a certain line of desks and chairs. The desk requires 6 labor-hours and the chair 2. These satisfy the relation $6x + 2y \le 320$. If $x = 30$ desks, find y, the number of chairs that can be made.

Health application **46.** An adult should have at least 55 grams of protein. On a restricted diet of bread, peanut butter, and milk, the relation is $2b + 5p + 8m \ge 55$. If the patient eats $b = 4$ slices of bread, and $p = 3$ tablespoons of peanut butter, how many glasses of milk m must he drink?

CHAPTER 2 SUMMARY

Important Words and Phrases

addition property of equality (2.2)
addition property of inequalities (2.6)
combining like terms (2.1)
equivalent equations (2.2)
general linear equations (2.4)
linear inequality (2.6)
multiplication property of equality (2.3)
multiplication property of inequalities (2.6)
translating English into mathematics (2.5)

Important Properties

Let A, B, and C be algebraic expressions.

If $A = B$, then $A + C = B + C$

$$A - C = B - C$$

$$\left. \begin{array}{l} AC = BC \\[6pt] \dfrac{A}{C} = \dfrac{B}{C} \end{array} \right\} \quad (C \ne 0)$$

If $A < B$, then $A + C < B + C$

$$A - C < B - C$$

$$\left.\begin{array}{c} AC < BC \\ \dfrac{A}{C} < \dfrac{B}{C} \end{array}\right\} \quad (C \; positive)$$

$$\left.\begin{array}{c} AC > BC \\ \dfrac{A}{C} > \dfrac{B}{C} \end{array}\right\} \quad (C \; negative)$$

(Similarly this is true for $\leq$, $>$, and $\geq$.)

Important Procedures

To solve linear equations and inequalities:
1. Use the distributive law to remove the parentheses.
2. Combine like terms.
3. Move all terms with the unknown to one side.
4. Move all number terms to the other side.
5. Multiply or divide to isolate the unknown.
6. Check the solution.

CHAPTER 2 REVIEW EXERCISES

Simplify the following expressions as much as possible.

1. $7(3x + 2)$ 2. $5a - 7 - 8 - 3$

3. $4u - 5u - 3u + 2$ 4. $6t - 10 - 3t - 3 + 2t$

5. $5(2 - 3t) - 2(4t - 1)$

Are the following equations equivalent?

6. $x = 5$; $3x = 15$; $2x + 1 = 11$; $12 - x = x + 2$

Solve the following equations and inequalities. (Graph the solutions to the inequalities.)

7. $x - 7 = 6$ 8. $a + 5 = 2$

9. $10t - 5 - 9t = 6$

10. $8t - 5 = 7t + 3$

11. $3x = -18$

12. $\dfrac{x}{5} = 4$

13. $\dfrac{-2}{9}x = 20$

14. $-t = 8$

15. $3x + 2x - 1 = 9$

16. $6u - 3 = 8u + 5$

17. $3(k + 5) = 8k - 5$

18. $3(5 - a) = 4(9 + a)$

19. $5(x + 3) + 3x = 6(x + 4) - 5$

20. $4(10 - a) - 3(a + 3) = -9(2a - 5) + 19$

21. $x + 5 < 18$

22. $a - 6 \leq 4$

23. $5k \geq -25$

24. $\dfrac{-x}{3} > 21$

25. $6x - 17 < 2x + 23$

26. $3x - 4(2 + 5x) \geq 43$

Solve the following word problems.

27. One number is 1.5 times another number. Their sum is 20. Find the numbers.

28. Seventy-five is 5% of what number?

29. Mary has 20 pieces of candy to give to her two children. Jason gets 2 pieces more than Jennifer. How many pieces does each child get?

30. The length of a rectangle is 3 more than twice the width. The perimeter is 42. Find the length and width.

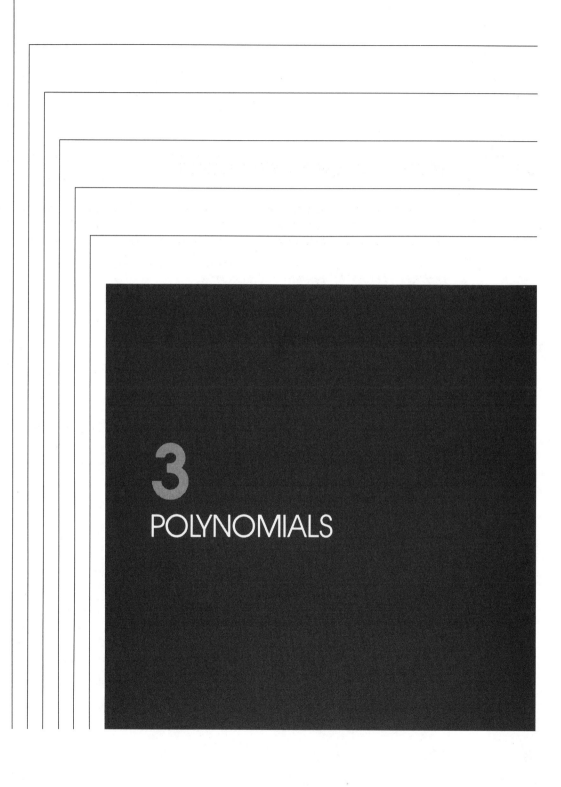

3
POLYNOMIALS

3.1   INTRODUCTION TO EXPONENTS

We recall that repeated addition, such as $8 + 8 + 8 + 8 + 8$, can be written as multiplication $5 \cdot 8$. Similarly, repeated multiplication, such as $7 \cdot 7 \cdot 7 \cdot 7$, can be written in **exponent form** 7^4. In general, we have the following notation.

DEFINITION 1   For any real number x,

$$x^n = \underbrace{x \cdot x \cdot x \cdot \ldots \cdot x}_{n\text{-factors}}$$

Here x is called the **base**, and n is called the **exponent** or **power**.

EXAMPLE 1   The following are examples of exponents.

(a) $x^2 = x \cdot x$ (read "x squared")   **(b)** $5^3 = 5 \cdot 5 \cdot 5 = 125$ (read "5 cubed")

(c) $a^4 = a \cdot a \cdot a \cdot a$
(read "a to the fourth power")   **(d)** $2^7 = 2 \cdot 2 \cdot 2 \cdot 2 \cdot 2 \cdot 2 \cdot 2 = 128$
(read "2 to the seventh power")

(e) $t^1 = t$ **(f)** $(-4)^2 = (-4)(-4) = 16$

(g) $(0.7)^3 = (0.7)(0.7)(0.7) = 0.343$ **(h)** $(-10)^3 = (-10)(-10)(-10) = -1000$

(i) $\left(\dfrac{2}{3}\right)^4 = \left(\dfrac{2}{3}\right)\left(\dfrac{2}{3}\right)\left(\dfrac{2}{3}\right)\left(\dfrac{2}{3}\right) = \dfrac{16}{81}$

YES	NO
$(-3)^2 = (-3)(-3) = 9$	$(-3)^2 = -9$
$-2^4 = -(2^4) = -16$	$-2^4 = (-2)^4 = 16$

The negative sign within the parentheses is a part of the base. When there are no parentheses, the negative sign is *not* a part of the base.

Consider what happens when we multiply numbers such as $x^3 \cdot x^4$.

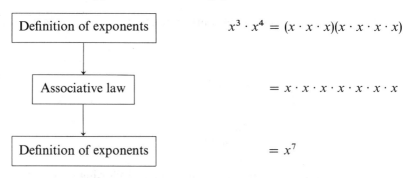

$$x^3 \cdot x^4 = (x \cdot x \cdot x)(x \cdot x \cdot x \cdot x)$$

$$= x \cdot x \cdot x \cdot x \cdot x \cdot x \cdot x$$

$$= x^7$$

Since $3 + 4 = 7$, this suggests the following property.

PROPERTY 1   For any real number a and positive integers n and m,

$$a^n \cdot a^m = a^{n+m}$$

EXAMPLE 2   The following examples use Property 1.

(a) $a^7 \cdot a^6 = a^{7+6} = a^{13}$

(b) $x \cdot x^5 = x^{1+5} = x^6$

(c) $4^3 \cdot 4^5 = 4^{3+5} = 4^8$

(d) $(m + n)^4(m + n)^{11} = (m + n)^{4+11} = (m + n)^{15}$

(e) $t^2 \cdot t^3 \cdot t^4 = t^{2+3+4} = t^9$

YES	NO
$x^8 \cdot x^4 = x^{12}$	~~$x^8 \cdot x^4 = x^{32}$~~

We *add* exponents when we multiply numbers in exponent form.

YES	NO
$2^5 \cdot 3^7 = 2^5 \cdot 3^7$	~~$2^5 \cdot 3^7 = 6^{12}$~~

The bases must be the *same* to use Property 1.

Now consider $(a^3)^4$.

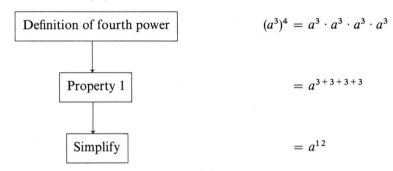

Definition of fourth power	$(a^3)^4 = a^3 \cdot a^3 \cdot a^3 \cdot a^3$
Property 1	$= a^{3+3+3+3}$
Simplify	$= a^{12}$

Since $3 \cdot 4 = 12$, this suggests the following property.

PROPERTY 2   For any real number a and positive integers m and n,

$$(a^m)^n = a^{m \cdot n}$$

EXAMPLE 3   The following examples use Property 2.

(a) $(x^6)^2 = x^{6 \cdot 2} = x^{12}$

(b) $(5^3)^7 = 5^{3 \cdot 7} = 5^{21}$

(c) $(t^4)^9 = t^{4 \cdot 9} = t^{36}$

(d) $(10^3)^2 = 10^{3 \cdot 2} = 10^6$

YES	NO
$(10^3)^2 = 10^6$	$(10^3)^2 = 10^9$

We *multiply* the exponents when we use Property 2.

Now consider $(3a)^4$.

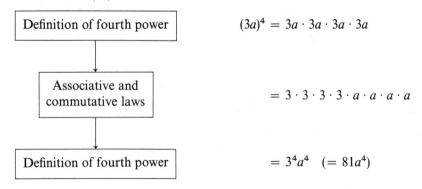

Definition of fourth power	$(3a)^4 = 3a \cdot 3a \cdot 3a \cdot 3a$
Associative and commutative laws	$= 3 \cdot 3 \cdot 3 \cdot 3 \cdot a \cdot a \cdot a \cdot a$
Definition of fourth power	$= 3^4 a^4 \quad (= 81a^4)$

This suggests the following property.

PROPERTY 3   For any real numbers a and b and positive integer k,

$$(ab)^k = a^k b^k$$

EXAMPLE 4   The following examples use Property 3.

    (a) $(7x)^3 = 7^3 x^3 = 343x^3$

    (b) $(2abc)^4 = 2^4 a^4 b^4 c^4 = 16a^4 b^4 c^4$

    (c) $(3mn^2)^4 = 3^4 m^4 (n^2)^4 = 81m^4 n^8$

PROBLEM SET 3.1

Write the following expressions in exponent form.

See Example 1

1. $x \cdot x \cdot x \cdot x \cdot x \cdot x$

2. $7 \cdot 7 \cdot 7 \cdot 7 \cdot 7$

3. $2 \cdot 2 \cdot 2 \cdot 2 \cdot 2 \cdot 2 \cdot 2 \cdot 2$

4. $a \cdot a \cdot a$

5. $(p + q)(p + q)$

6. $\left(\dfrac{3}{4}\right)\left(\dfrac{3}{4}\right)\left(\dfrac{3}{4}\right)\left(\dfrac{3}{4}\right)\left(\dfrac{3}{4}\right)$

Write each of the following as a single number.

See Example 1

7. 2^4

8. 8^2

9. 10^3

10. 3^5

11. $(-4)^3$

12. $(-7)^2$

13. $(-10)^4$

14. $(1.7)^2$

15. $(0.6)^2$

16. $\left(\dfrac{1}{3}\right)^4$

17. $\left(\dfrac{-2}{5}\right)^2$

18. $\left(\dfrac{-1}{10}\right)^5$

Simplify each of the following as much as possible.

See Example 2

19. $x^2 x^3$

20. $a^5 a^9$

21. $3^4 \cdot 3^5$

22. $2^3 \cdot 2^{10}$

23. $(a + b)^3 (a + b)^4$

24. $(p - q)^4 (p - q)^7$

25. $a^3 a^4 a^5$

26. $b^7 b^2 b$

27. $10^2 \cdot 10^3 \cdot 10^4$ **28.** $4^5 \cdot 4^7 \cdot 4^3$

See   **29.** $(x^2)^5$ **30.** $(a^3)^4$
Example 3
          31. $(z^4)^4$ **32.** $(b^5)^3$

          33. $(8^3)^7$ **34.** $(10^2)^3$

          35. $(9^2)^4$ **36.** $(5^8)^2$

See   **37.** $(3x)^3$ **38.** $(5a)^2$
Example 4
          39. $(4ab)^2$ **40.** $(2xy)^5$

          41. $(3abc)^4$ **42.** $(10mn)^3$

          43. $(5ab^2)^3$ **44.** $(6pq^3)^2$

Life science **45.** If a man and a woman each have 2^{23} possible combinations of the 23
application   chromosomes, together they have $2^{23} \cdot 2^{23}$ combinations. Write this as
                    a single exponent.

Business **46.** If 8% annual interest is compounded quarterly, after 1 year the previous
application   balance is multiplied by $(1.02)^4$. After 6 years, this is $[(1.02)^4]^6$. Rewrite
                 this with a single exponent.

Science **47.** The speed of light is given by $c = 3 \cdot 10^8$ meters per second. In the theory
application   of relativity, the quantity c^2 is used. Compute c^2.

3.2   MORE EXPONENT RULES

In this section we consider division problems, such as $\dfrac{x^7}{x^2}$.

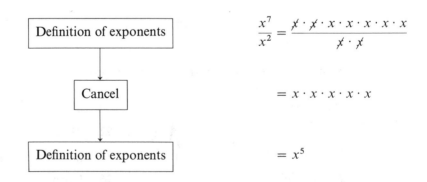

Since $5 = 7 - 2$, this suggests that we subtract exponents to divide. Now consider $\dfrac{x^3}{x^5}$.

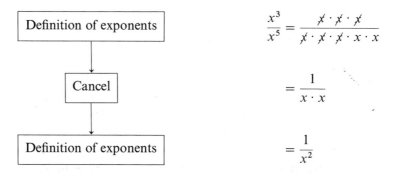

$$\frac{x^3}{x^5} = \frac{\cancel{x} \cdot \cancel{x} \cdot \cancel{x}}{\cancel{x} \cdot \cancel{x} \cdot \cancel{x} \cdot x \cdot x}$$

$$= \frac{1}{x \cdot x}$$

$$= \frac{1}{x^2}$$

This suggests the following property.

PROPERTY 4 For any real number a ($\neq 0$) and positive integers m and n,

$$\frac{a^m}{a^n} = \begin{cases} a^{m-n} & \text{if } m > n \\ 1 & \text{if } m = n \\ \dfrac{1}{a^{n-m}} & \text{if } n > m \end{cases}$$

In words, if the numerator has the larger exponent, the result is in the numerator; if the denominator has the larger exponent, the result is in the denominator.

EXAMPLE 5 The following examples use Property 4.

(a) $\dfrac{x^7}{x^5} = x^{7-5} = x^2$

(b) $\dfrac{2^{10}}{2^3} = 2^{10-3} = 2^7$

(c) $\dfrac{t^5}{t^5} = 1$

(d) $\dfrac{a^4}{a^9} = \dfrac{1}{a^{9-4}} = \dfrac{1}{a^5}$

(e) $\dfrac{3^2}{3^8} = \dfrac{1}{3^{8-2}} = \dfrac{1}{3^6}$

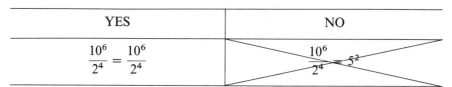

YES	NO
$\dfrac{x^{12}}{x^3} = x^9$	$\dfrac{x^{12}}{x^3} = x^4$

We *subtract* exponents when we divide.

YES	NO
$\dfrac{10^6}{2^4} = \dfrac{10^6}{2^4}$	$\dfrac{10^6}{2^4} = 5^2$

Like Property 1, we must have the *same* bases to use Property 4.

The last case that we consider is $\left(\dfrac{x}{y}\right)^5$.

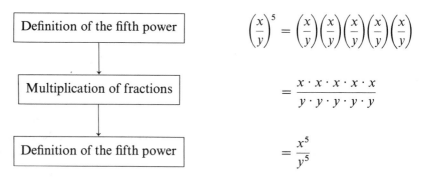

Definition of the fifth power	$\left(\dfrac{x}{y}\right)^5 = \left(\dfrac{x}{y}\right)\left(\dfrac{x}{y}\right)\left(\dfrac{x}{y}\right)\left(\dfrac{x}{y}\right)\left(\dfrac{x}{y}\right)$
Multiplication of fractions	$= \dfrac{x \cdot x \cdot x \cdot x \cdot x}{y \cdot y \cdot y \cdot y \cdot y}$
Definition of the fifth power	$= \dfrac{x^5}{y^5}$

This suggests our final property.

PROPERTY 5   For any real numbers a and b ($b \neq 0$) and positive integer k,

$$\left(\frac{a}{b}\right)^k = \frac{a^k}{b^k}$$

EXAMPLE 6   The following examples use Property 5.

(a) $\left(\dfrac{2}{x}\right)^4 = \dfrac{2^4}{x^4} = \dfrac{16}{x^4}$

(b) $\left(\dfrac{m}{n}\right)^7 = \dfrac{m^7}{n^7}$

(c) $\left(\dfrac{ab}{3}\right)^2 = \dfrac{a^2b^2}{9}$

(d) $\left(\dfrac{p^3q^4}{r^2}\right)^3 = \dfrac{(p^3)^3(q^4)^3}{(r^2)^3} = \dfrac{p^9q^{12}}{r^6}$

Consider the problem $\dfrac{x^4}{x^6}$. We can work this two ways.

USING PROPERTY 4	SUBTRACTING EXPONENTS
$\dfrac{x^4}{x^6} = \dfrac{1}{x^2}$	$\dfrac{x^4}{x^6} = x^{4-6} = x^{-2}$

We want these two to be the same. Therefore, we make the following definition for **negative exponents.**

DEFINITION 2   For any real number a ($\neq 0$) and positive integer n,

$$a^{-n} = \dfrac{1}{a^n}$$

Similarly, we want $\dfrac{x^5}{x^5}$ to be 1 and also $x^{5-5} = x^0$. Thus, for any real number a ($a \neq 0$), we have a **zero exponent.**

DEFINITION 3   For any real number a ($\neq 0$),

$$a^0 = 1$$

EXAMPLE 7   The following are examples of the definitions of negative and zero exponents.

(a) $x^{-3} = \dfrac{1}{x^3}$

(b) $a^{-5} = \dfrac{1}{a^5}$

(c) $2^{-4} = \dfrac{1}{2^4} = \dfrac{1}{16}$

(d) $3^{-1} = \dfrac{1}{3^1} = \dfrac{1}{3}$

(e) $6^0 = 1$

(f) $t^0 = 1$

YES	NO
$5^{-2} = \dfrac{1}{5^2} = \dfrac{1}{25}$	~~$5^{-2} = 5^2 = -25$~~
$3^0 = 1$	~~$3^0 = 0$~~

With these new definitions, Property 4 can be rewritten simply as $\dfrac{a^m}{a^n} = a^{m-n}$. Let us summarize all the definitions and properties of exponents.

Summary of Exponent Rules   For any real numbers a and b and positive integers $m, n, k,$

$$a^n = a \cdot a \cdot a \cdot \ldots \cdot a \quad (n\text{-factors})$$

$$a^0 = 1 \quad (a \neq 0)$$

$$a^{-n} = \frac{1}{a^n} \quad (a \neq 0)$$

$$a^n a^m = a^{n+m}$$

$$(a^m)^n = a^{mn}$$

$$(ab)^k = a^k b^k$$

$$\left(\frac{a}{b}\right)^k = \frac{a^k}{b^k} \quad (b \neq 0)$$

$$\frac{a^m}{a^n} = \begin{cases} a^{m-n} & \text{if } m > n \quad (a \neq 0) \\ 1 & \text{if } m = n \quad (a \neq 0) \\ \dfrac{1}{a^{n-m}} & \text{if } n > m \quad (a \neq 0) \end{cases}$$

$$\frac{a^m}{a^n} = a^{m-n} \quad (a \neq 0)$$

PROBLEM SET 3.2

Simplify the following terms. Leave all answers with positive exponents.

<table>
<tr><td rowspan="2">See
Example 5</td><td>1. $\dfrac{x^{10}}{x^6}$</td><td>2. $\dfrac{a^{12}}{a^{10}}$</td><td>3. $\dfrac{t^7}{t^3}$</td></tr>
<tr><td>4. $\dfrac{5^6}{5^4}$</td><td>5. $\dfrac{2^6}{2^6}$</td><td>6. $\dfrac{10^3}{10^7}$</td></tr>
</table>

7. $\dfrac{z^3}{z^3}$ **8.** $\dfrac{k^9}{k^8}$ **9.** $\dfrac{m^{11}}{m^9}$

10. $\dfrac{9^5}{9^8}$ **11.** $\dfrac{3^2}{3^8}$ **12.** $\dfrac{6^{10}}{6^7}$

See
Example 6

13. $\left(\dfrac{3}{a}\right)^2$ **14.** $\left(\dfrac{x}{y}\right)^4$ **15.** $\left(\dfrac{m}{2}\right)^3$

16. $\left(\dfrac{t}{s}\right)^{10}$ **17.** $\left(\dfrac{2r}{p}\right)^4$ **18.** $\left(\dfrac{abc}{2}\right)^5$

19. $\left(\dfrac{xy}{z}\right)^3$ **20.** $\left(\dfrac{5}{uv}\right)^2$ **21.** $\left(\dfrac{x^2y}{z}\right)^4$

22. $\left(\dfrac{ab^3}{c^2}\right)^3$ **23.** $\left(\dfrac{2mk^2}{5n^3}\right)^3$ **24.** $\left(\dfrac{3x^2y}{4z^3}\right)^2$

See
Example 7

25. x^{-6} **26.** a^{-7} **27.** 2^{-5}

28. 3^{-3} **29.** 5^{-2} **30.** 4^{-3}

31. 10^{-4} **32.** $\left(\dfrac{1}{2}\right)^{-2}$ **33.** r^0

34. t^0 **35.** 7^0 **36.** $(-4)^0$

Business
application

37. At 9% interest, $1000 eight years from now has a present value of $1000(1.09)^{-8}$. Write this with a positive exponent.

Health
application

38. There is a 19-in-20 chance that a patient will not experience a harmful side effect from a certain drug. For 12 patients, the chance that none of them will experience a harmful side effect is

$$\left(\frac{19}{20}\right)^{12}$$

Rewrite this as a ratio of exponent terms.

Consumer
application
39. The likelihood of winning a certain million-dollar lottery is 10^{-7}. Rewrite this as a fraction.

3.3   MONOMIALS

A **monomial** (or **term**) is the product of a constant and a variable and/or powers of variables. The numerical constant is called the **coefficient.**

EXAMPLE 8   The following are examples of monomials and their coefficients.

	MONOMIAL	COEFFICIENT
(a)	$2ab$	2
(b)	$-4x^2yz^3$	-4
(c)	mnk^3	1
(d)	$-p^5q^2$	-1

Notice from examples **(c)** and **(d)** that the coefficient is understood to be 1 (or -1) if no numeral appears, since $mnk^3 = 1 \cdot mnk^3$ and $-p^5q^2 = (-1)p^5q^2$.

As we saw in Section 2.1, monomials with the same variable part are called **like terms.** Recall that we use the distributive law to combine like terms.

> To add (or subtract) like terms or monomials, use the distributive law to add (or subtract) the coefficients.

EXAMPLE 9   We combine the coefficients to combine like monomials.

(a) $2x^2y + 3x^2y + 4x^2y = (2 + 3 + 4)x^2y = 9x^2y$

(b) $4abc - abc = (4 - 1)abc = 3abc$

(c) $15p^2q^3 + 4p^2q^3 - p^2q^3 = (15 + 4 - 1)p^2q^3 = 18p^2q^3$

YES	NO
$2a + 3b = 2a + 3b$	$2a + 3b = 5ab$

Combining $2a$ and $3b$ is like adding apples and oranges. We must have like terms to combine.

EXAMPLE 10   Multiply $(2ab^2)(5a^3b^4)$.

SOLUTION   We multiply the coefficients and the variables of like base separately.

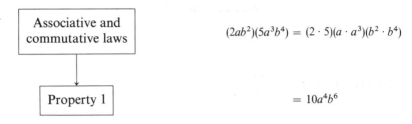

$$(2ab^2)(5a^3b^4) = (2 \cdot 5)(a \cdot a^3)(b^2 \cdot b^4)$$

$$= 10a^4b^6$$

This example suggests the following rule.

> To multiply monomials:
>
> **1.** Multiply the coefficients.
> **2.** Multiply the variables of like base (add the exponents with the same base).

EXAMPLE 11   Multiply $(3xy^2)(-2x^2y^5z)(5x^3y^3z^6)$.

SOLUTION   We multiply the coefficients and the variables of like base separately.

Multiply coefficients; add exponents

$$(3xy^2)(-2x^2y^5z)(5x^3y^3z^6) = [3(-2)5][xx^2x^3][y^2y^5y^3][zz^6]$$
$$= -30x^6y^{10}z^7$$

Dividing monomials is very similar.

> To divide monomials:
>
> **1.** Divide the coefficients.
> **2.** Use exponent properties to divide variables of like base.

EXAMPLE 12   Divide $\dfrac{12mn^4}{2m^3n^3}$.

SOLUTION   As with multiplication, we divide the coefficients and variables of like base separately.

$$\frac{12mn^4}{2m^3n^3} = \left[\frac{12}{2}\right]\left[\frac{m}{m^3}\right]\left[\frac{n^4}{n^3}\right]$$

> Divide coefficients;
> subtract exponents

$$= \frac{6n}{m^2}$$

EXAMPLE 13   Divide $\dfrac{15a^2b^3c^4}{-3a^4b^3c^2d}$.

SOLUTION   We divide the coefficients and variables of like base separately.

$$\frac{15a^2b^3c^4}{-3a^4b^3c^2d} = \left[\frac{15}{-3}\right]\left[\frac{a^2}{a^4}\right]\left[\frac{b^3}{b^3}\right]\left[\frac{c^4}{c^2}\right]\left[\frac{1}{d}\right]$$

> Divide coefficients;
> subtract exponents

$$= \frac{-5c^2}{a^2d}$$

PROBLEM SET 3.3

For each of the following monomials, give its coefficient.

*See
Example 8*

1. $5abc$ **2.** $-4x^2$ **3.** $10p^2q^3$

4. s^3t^5 **5.** $-m$ **6.** xyz

Simplify the following expressions.

*See
Example 9*

7. $3x + 4x - 5x$ **8.** $7a^2b + 9a^2b$

9. $pq - 7pq + 3pq$ **10.** $6m^2k - 10m^2k - m^2k$

11. $3xy + 6xy - 10xy + 4xy$ **12.** $10t^2 - 3t^2 + 8t^2$

13. $abcd - 9abcd$ **14.** $3uv^3 - 11uv^3 - 5uv^3$

Multiply the following monomials.

*See
Examples
10 and 11*

15. $(3x)(4x^2)$ **16.** $(7a^2)(8a^3)$

17. $(-8r^4)(3r)$ **18.** $(-4m^5)(-2m^2)$

19. $(5ab^2)(3a^2b)$ **20.** $(7pq)(-4p^2q)$

21. $(4x^2yz^3)(-6xy^4z^2)$ **22.** $(9ab^2c^3)(-2a^2b^3c)$

23. $(10uv^2)(3u^2v^5)(2u^7v)$ **24.** $(-x^2y^5)(-2xy^7)(3x^2y^3)$

25. $(-ab^2c^3)(5a^2b)(4b^3c)$ **26.** $(4r^2s^3)(3s^2t)(-2r^4t^5)$

Divide the following monomials.

See Examples 12 and 13

27. $\dfrac{6x^5}{3x^2}$ **28.** $\dfrac{10a^7}{5a^2}$

29. $\dfrac{-12u^4}{3u^5}$ **30.** $\dfrac{-15k^7}{-k}$

31. $\dfrac{24ab^3}{-2a^2b^2}$ **32.** $\dfrac{16x^2y^5}{8x^3y^3}$

33. $\dfrac{20u^2v^3}{-10uv}$ **34.** $\dfrac{30rs^2t^3}{15r^2s^2t^2}$

35. $\dfrac{5xy^2z^4w^6}{10x^4yz^4}$ **36.** $\dfrac{8a^2bc^3d}{-2ab^2c^3d^4e^5}$

37. $\dfrac{-9m^2n^3k^5}{-3m^4nk^4j^5}$ **38.** $\dfrac{-10u^2v^3w^4x^5}{30uv^3w^5x^7}$

3.4 ADDING AND SUBTRACTING POLYNOMIALS

A **polynomial** is a sum of monomials (or terms). For example, $2x + 3xy + 7x^5$ is a polynomial (the sum of three monomials). Some polynomials have special names. One term is a **monomial.** The sum of two monomials is a **binomial.** The sum of three monomials is a **trinomial.**

If a polynomial has only one variable, the **degree** of the polynomial is the highest power to which the variable is raised.

EXAMPLE 14   The following are various polynomials with their degrees and special names (if any).

POLYNOMIAL	DEGREE	SPECIAL NAME
(a) $x^5 + 2x$	5	Binomial
(b) $10a^3$	3	Monomial
(c) $u^4 - u^3 - 6u$	4	Trinomial
(d) $1 + m^2 + m^4 + m^6 + m^8$	8	Polynomial

In Chapter 2 we saw how to simplify expressions by combining like terms. Adding polynomials is very similar.

> **To add polynomials:**
>
> **1.** Group (or line up vertically) all like terms.
> **2.** Add the coefficients. (Combine like terms.)

EXAMPLE 15   Add $(4x^2 - 2x + 3) + (3x^3 + 4x - 1) + (2x^3 - 3x^2 + 10)$.

SOLUTION   We can add polynomials horizontally by grouping or vertically by lining up common terms. We work this problem vertically (and the next one horizontally). Notice that we leave spaces for the missing powers of x.

Line up like terms

↓

Add coefficients

$$
\begin{array}{r}
4x^2 - 2x + 3 \\
3x^3 + 4x - 1 \\
2x^3 - 3x^2 + 10 \\
\hline
5x^3 + x^2 + 2x + 12
\end{array}
$$

EXAMPLE 16   Add $(6u^3 - u + 5) + (4u^2 + 8u - 3) + (5u^3 - u^2 - u)$.

SOLUTION   We add these polynomials horizontally by grouping the like terms together.

Given

$(6u^3 - u + 5) + (4u^2 + 8u - 3) + (5u^3 - u^2 - u)$

↓

Group like terms

$= (6u^3 + 5u^3) + (4u^2 - u^2) + (-u + 8u - u) + (5 - 3)$

↓

Simplify

$= 11u^3 + 3u^2 + 6u + 2$

Recall that subtracting real numbers meant adding the opposite. The same is true for polynomials. The **opposite of a polynomial** is the polynomial we get by *changing all the signs*. For instance, the opposite of $6x^2 - 4x + 5$ is $-6x^2 + 4x - 5$.

> To subtract polynomials:
>
> **1.** Change *all* the signs in the polynomial to be subtracted. (Find its opposite.)
>
> **2.** Add the resulting polynomials.

EXAMPLE 17   Find $D = (6x^3 - 2x^2 - 7x + 8) - (5x^2 - 6x + 4)$.

SOLUTION   We change *all* the signs of the second polynomial (this is now the opposite), and then we add.

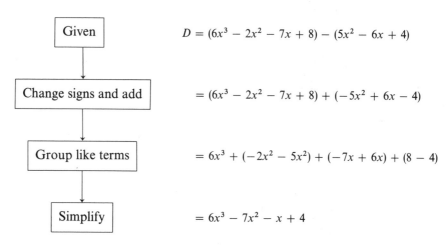

| Given | $D = (6x^3 - 2x^2 - 7x + 8) - (5x^2 - 6x + 4)$ |

Change signs and add $= (6x^3 - 2x^2 - 7x + 8) + (-5x^2 + 6x - 4)$

Group like terms $= 6x^3 + (-2x^2 - 5x^2) + (-7x + 6x) + (8 - 4)$

Simplify $= 6x^3 - 7x^2 - x + 4$

EXAMPLE 18   Simplify $(4a^3 - 7a^2 - 3) + (5a^3 - 8a + 10) - (-a^2 + 5a + 2)$.

SOLUTION   We replace the last term by its opposite by changing all of its signs. We then add these vertically.

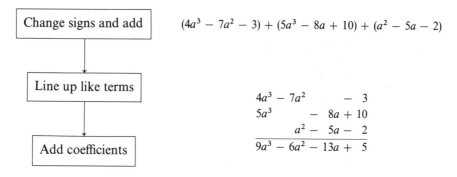

Change signs and add $(4a^3 - 7a^2 - 3) + (5a^3 - 8a + 10) + (a^2 - 5a - 2)$

Line up like terms

Add coefficients

$$
\begin{array}{rrrr}
4a^3 & -\ 7a^2 &   & -\ 3 \\
5a^3 &   & -\ 8a & +\ 10 \\
     & a^2   & -\ 5a & -\ 2 \\
\hline
9a^3 & -\ 6a^2 & -\ 13a & +\ 5
\end{array}
$$

PROBLEM SET 3.4

For each of the following polynomials, give the degree and its special name (if any).

See Example 14

1. $2x^2 + 5x + 1$ **2.** $4x^5 + 3x^2$

3. $a^2 - 4a^3$ **4.** $-10u^4$

5. $1 + t + t^2 + t^3 + t^4$ **6.** $m^4 - m^3 - m^2$

7. $4z$ **8.** $3 - 4v^2 + 5v^4 - v^7 + v^{12}$

Add the following polynomials.

See Example 15

9. $\begin{aligned} 3x^2 + 5x - 1 \\ 4x^2 + 7x + 3 \end{aligned}$ **10.** $\begin{aligned} 7a^2 - 9a + 1 \\ 4a^2 + 5a - 3 \end{aligned}$

11. $\begin{aligned} 4u^4 + 3u^3 \quad\;\; - u + 7 \\ 5u^4 \quad\quad\;\; + 3u^2 + 7u - 3 \end{aligned}$ **12.** $\begin{aligned} 10x^4 \quad\quad\; + 6x^2 - 7x + 2 \\ 2x^4 - 3x^3 - 8x^2 \quad\quad\; + 5 \end{aligned}$

13. $\begin{aligned} 8t^4 - 3t^3 - \;\; t^2 + \;\; t + 1 \\ 5t^4 \quad\quad\;\; + 2t^2 + 5t + 3 \\ 8t^3 + 6t^2 \quad\quad\; - 7 \end{aligned}$

14. $\begin{aligned} 12m^4 - \;\; 8m^3 \quad\quad\; + 4m - 5 \\ 6m^4 \quad\quad\;\; + 7m^2 - 5m + 7 \\ 10m^3 - 3m^2 + 6m - 8 \end{aligned}$

See Example 16

15. $(4x^2 - 8x + 7) + (5x^2 + 6x - 3)$

16. $(8v^2 + 4v - 9) + (3v^2 - 7v + 1)$

17. $(4r^3 - 3r^2 + 5r + 1) + (7r^3 + 3r^2 - 6r + 3)$

18. $(8y^3 - 2y^2 + 5y + 1) + (6y^3 - 5y^2 + 7y - 8)$

19. $(10t^3 - 5t + 2) + (6t^3 + 9t^2 - 5) + (7t^2 - 4t - 12)$

20. $(4m^4 - 3m^3 + 5m + 7) + (6m^3 + 5m^2 - 2) + (m^3 - m^2 + m + 1)$

Subtract the following polynomials.

See Example 17

21. $(4x^2 - 3x + 1) - (3x^2 + 5x - 2)$

22. $(8u^2 - 7u - 5) - (6u^2 + 5u - 6)$

23. $(10m^3 + 3m^2 - 5m + 2) - (6m^3 - 4m^2 + 8m - 5)$

24. $(4s^3 - 3s^2 + 6s + 7) - (9s^3 - 8s^2 + 5s + 2)$

25. $(5t^3 - 4t + 6) - (8t^2 + 5t + 7)$

26. $(8z^3 - z^2 + 6) - (7z^2 - 10z + 4)$

27. $(3x^4 - 4x^2 - x + 1) - (x^3 + x^2 - 3x + 5)$

28. $(9a^5 - 12a^3 - 6a) - (4a^4 - a^3 - 6a^2 - 7)$

Simplify the following expressions.

See
Example 18

29. $(3x^2 + 5x) + (8x^2 - 3x) - (4x^2 + 4x)$

30. $(5a - 3) + (8a + 7) - (2a - 6)$

31. $(4x + 10) - (3x - 4) - (2x - 7)$

32. $(3u^2 - 7) - (4u^2 - 8) - (3u^2 + 8)$

33. $(4r^2 - r + 1) - (3r^2 - 3r + 5) + (5r^2 - r + 1)$

34. $(8a^2 + a + 6) + (7a^2 - 4a - 5) - (2a^2 - 3a + 1)$

35. $(10m^3 - 3m + 1) + (6m^2 - 8m + 7) - (4m^3 - m^2 + 7)$

36. $(5t^3 - 4t^2 - t + 1) - (8t^2 - 6t + 5) - (7t^3 - 3t^2 - t)$

Business
application

37. In business, the profit P is the revenue R less the cost C. As a formula, this is written

$$P = R - C$$

Use this formula to complete the following table.

Profit (P)	Revenue (R)	Cost (C)
?	$100x$	$x^2 + 25x + 2000$
?	$75x$	$0.5x^2 + 38x + 8000$

3.5   MULTIPLYING POLYNOMIALS

Before we begin multiplying general polynomials, we multiply a monomial by a polynomial using the distributive law.

EXAMPLE 19   Multiply $2ab^2(3a + 4b^2 - 5a^2b)$.

SOLUTION   Using the distributive law, we multiply the $2ab^2$ term by each of the terms within the parentheses. We multiply each of these as monomials.

$$\boxed{\text{Distributive law}} \qquad 2ab^2(3a + 4b^2 - 5a^2b) = 6a^2b^2 + 8ab^4 - 10a^3b^3$$

In general, to multiply polynomials, we use the distributive law twice.

$$(a + b)(c + d) = (a + b)c + (a + b)d = ac + bc + ad + bd$$

This suggests the following rule.

> To multiply two polynomials:
>
> **1.** Multiply all the terms of one polynomial by all the terms of the other.
> **2.** Combine like terms.

This procedure is very similar to multiplying whole numbers. Notice the similarity in the examples below.

$$
\begin{array}{r}
32 \\
\times\, 21 \\
\hline
32 \\
64 \\
\hline
672
\end{array}
\qquad
\begin{array}{r}
3a + 2b \\
\times\ 2a +\ b \\
\hline
3ab + 2b^2 \\
6a^2 + 4ab \\
\hline
6a^2 + 7ab + 2b^2
\end{array}
$$

EXAMPLE 20   Multiply $(3x + 5)(2x - 3)$.

SOLUTION   We multiply this out vertically, multiplying every term of the first polynomial by every term of the second.

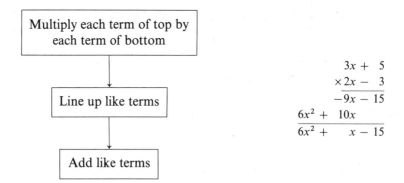

$$
\begin{array}{r}
3x +\ 5 \\
\times\, 2x -\ 3 \\
\hline
-9x - 15 \\
6x^2 +\ 10x \\
\hline
6x^2 +\quad x - 15
\end{array}
$$

EXAMPLE 21   Multiply $(2a - 5)(6a^2 - 3a + 4)$.

SOLUTION   We multiply this out vertically.

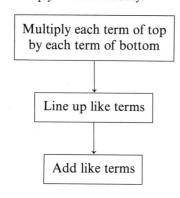

$$
\begin{array}{r}
6a^2 - 3a + 4 \\
\times \quad\quad 2a - 5 \\
\hline
-30a^2 + 15a - 20 \\
12a^3 - 6a^2 + 8a \\
\hline
12a^3 - 36a^2 + 23a - 20
\end{array}
$$

EXAMPLE 22   Multiply $(m^2 - 3m + 2)(m^3 - 8m^2 + 4m - 7)$.

SOLUTION   Again, we set this up as we would set up the multiplication of whole numbers.

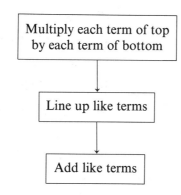

$$
\begin{array}{r}
m^3 - 8m^2 + 4m - 7 \\
\times \quad\quad m^2 - 3m + 2 \\
\hline
2m^3 - 16m^2 + 8m - 14 \\
-3m^4 + 24m^3 - 12m^2 + 21m \\
m^5 - 8m^4 + 4m^3 - 7m^2 \\
\hline
m^5 - 11m^4 + 30m^3 - 35m^2 + 29m - 14
\end{array}
$$

EXAMPLE 23   Multiply $(3x - 4)(2x + 3)(x - 5)$.

SOLUTION   As with real numbers, we multiply polynomials two at a time. We multiply the first two, and then multiply the product by the third.

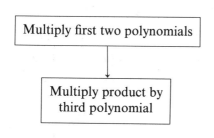

$$
\begin{array}{r}
3x - 4 \\
\times 2x + 3 \\
\hline
9x - 12 \\
6x^2 - 8x \\
\hline
6x^2 + x - 12 \\
\times \quad\quad x - 5 \\
\hline
-30x^2 - 5x + 60 \\
6x^3 + x^2 - 12x \\
\hline
6x^3 - 29x^2 - 17x + 60
\end{array}
$$

PROBLEM SET 3.5

Multiply the following polynomials.

See
Example 19

1. $3x(5x^2 + 2x - 6)$

2. $6a(2 - 3a - 4a^2)$

3. $4ab(a^2 + 2ab + b^2)$

4. $3xy(x^2 + 2xy - y^2)$

5. $7p(8pq^2 - 2q - 4p^3)$

6. $-2u(u - 4uv - 6uv^2)$

7. $3s^2t(s - 4st^2 + 6s^3t)$

8. $mn^2(m - 4m^2n + 3m^3n^2)$

See
Example 20

9. $(3x - 1)(2x + 5)$

10. $(2a - 3)(2a - 4)$

11. $(u + 7)(4u - 3)$

12. $(6m - 1)(5m + 3)$

13. $(4t + 6)(2t + 5)$

14. $(3y - 8)(4y + 3)$

15. $(4k - 5)(8k - 1)$

16. $(5p + 8)(6p - 7)$

17. $(a - 3b)(2a - b)$

18. $(2u + 3v)(4u - 5v)$

19. $(4m - 3n)(5m + 6n)$

20. $(8p - q)(7p + q)$

21. $(4r - s)(3r + 2s)$

22. $(10y - 3z)(y + 2z)$

See
Example 21

23. $(x + 5)(x^2 - 2x - 3)$

24. $(a - 2)(a^2 - 3a + 5)$

25. $(2u + 3)(u^2 + 5u - 1)$

26. $(3r - 2)(2r^2 - 4r + 5)$

27. $(a + 2b)(a^2 - ab + b^2)$

28. $(x - 3y)(4x^2 - 3xy + 5y^2)$

29. $(2r + 3t)(4r^2 - 5rt + 6t^2)$

30. $(5m - 2n)(3m^2 - 4mn + n^2)$

See
Example 22

31. $(2x^2 - x + 5)(3x^3 - x^2 + 5x - 2)$

32. $(4a^2 - 2a + 3)(5a^3 + a^2 - 3a + 7)$

33. $(5u^2 - u + 3)(6u^4 - 3u^3 + 5u^2 - 2u + 3)$

34. $(7t^2 - 4)(5t^5 - 4t^4 - 6t^3 - t^2 + 3t + 2)$

35. $(m^3 + 3m^2 - 4m + 2)(2m^3 - 4m^2 + 5m - 1)$

36. $(r^3 - 4r^2 - 6r + 3)(2r^3 - 7r^2 + 8r - 2)$

See
Example 23

37. $(x + 1)(x + 2)(x + 3)$

38. $(a + 2)(a - 3)(a + 4)$

39. $(u - 3)(u - 5)(u + 2)$

40. $(r + 3)(2r - 1)(r - 4)$

41. $(x + 1)^2(x - 7)$

42. $(y - 2)^2(y + 5)$

43. $(2t + 1)(t - 5)(t^2 - 2t - 1)$

44. $(3z - 2)(z + 6)(2z^2 - z + 6)$

Engineering application

45. The deflection y in a beam held in a wall is given by

$$y = (L - x)(L - x)(11x - 21)$$

where L is the length of the beam and x is the distance from the wall. Write this expression as a polynomial.

3.6   SPECIAL BINOMIAL PRODUCTS

In Section 3.5 we saw how to multiply any two polynomials. In this section we look at some special products that can be done very quickly (usually mentally).

We have a special rule for multiplying two binomials very quickly. It is called the **FOIL** method.

> **FOIL** means Firsts + Outers + Inners + Lasts

EXAMPLE 24   Multiply $(3x + 2)(2x - 5)$.

SOLUTION   We multiply both terms of the first binomials by both terms of the second in this order: firsts, outers, inners, lasts.

$$\boxed{\text{FOIL}} \qquad (3x + 2)(2x - 5) = 6x^2 - 15x + 4x - 10 = 6x^2 - 11x - 10$$

EXAMPLE 25   Multiply $(5a - 2b)(3a + 4b)$.

SOLUTION   We use the FOIL method.

$$\boxed{\text{FOIL}} \qquad (5a - 2b)(3a + 4b) = 15a^2 + 20ab - 6ab - 8b^2$$
$$= 15a^2 + 14ab - 8b^2$$

We usually do this multiplication in our head by combining the middle two terms.

YES	NO
$(5u - 4)(3u + 2) = 15u^2 - 2u - 8$	$(5u - 4)(3u + 2) = 15u^2 - 8$

Do not forget to multiply the middle terms.

Suppose that we want the square of a binomial, such as $(a + b)^2$. We can expand this using the FOIL method.

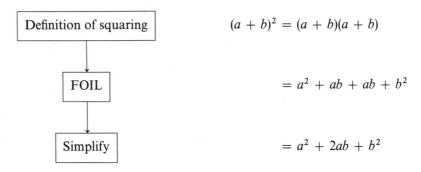

Definition of squaring	$(a + b)^2 = (a + b)(a + b)$
FOIL	$= a^2 + ab + ab + b^2$
Simplify	$= a^2 + 2ab + b^2$

This gives our properties for the **square of a binomial.**

> ***PROPERTY 6***   For any real numbers a and b,
>
> $$(a + b)^2 = a^2 + 2ab + b^2$$
> $$(a - b)^2 = a^2 - 2ab + b^2$$

EXAMPLE 26   The following are the squares of binomials and use the properties described above. These properties can be remembered as the rule: *Square of the first, twice the product, square of last.*

(a) $(x + 3)^2 = x^2 + 6x + 9$

(b) $(2a - 5)^2 = 4a^2 - 20a + 25$

(c) $(3u - 4v)^2 = 9u^2 - 24uv + 16v^2$

(d) $(10m + 5n)^2 = 100m^2 + 100mn + 25n^2$

square of first   twice product   square of last

YES	NO
$(a + 4)^2 = a^2 + 8a + 16$	$\cancel{(a + 4)^2 = a^2 + 16}$

We must add twice the product of the first and last terms.

Our last special product considers $(a + b)(a - b)$.

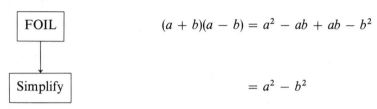

$$(a + b)(a - b) = a^2 - ab + ab - b^2$$

$$= a^2 - b^2$$

PROPERTY 7   For any real numbers a and b,

$$(a + b)(a - b) = a^2 - b^2$$

EXAMPLE 27   The following examples use the **sum and difference property.** Notice that we always end up with a difference of squares.

(a) $(s - t)(s + t) = s^2 - t^2$

(b) $(a - 6)(a + 6) = a^2 - 36$

(c) $(2p + 7q)(2p - 7q) = 4p^2 - 49q^2$

(d) $(1 - ab^2)(1 + ab^2) = 1 - a^2b^4$

PROBLEM SET 3.6

Multiply the following binomials mentally, using only the special-product rules.

See
Examples
24 and 25

1. $(x + 2)(x + 3)$

2. $(a - 2)(a + 5)$

3. $(u - 3)(u + 7)$

4. $(m - 5)(m - 8)$

5. $(a + 7)(2a + 1)$

6. $(3r - 5)(r + 6)$

7. $(2z - 7)(z + 8)$

8. $(4t - 1)(t + 3)$

9. $(3u - 2)(2u + 3)$

10. $(5x - 2)(3x + 4)$

11. $(6k + 3)(4k - 1)$

12. $(10z - 3)(2z + 5)$

13. $(7a + 2b)(3a - 5b)$

14. $(3x - 5y)(4x + 3y)$

15. $(5u - v)(7u + 3v)$

16. $(8p + 4q)(3p - 5q)$

17. $(2m - 5n)(3m + n)$

18. $(8r - s)(7r + s)$

See **19.** $(x + 4)^2$ **20.** $(a + 5)^2$
Example 26
21. $(r - 8)^2$ **22.** $(u - 2)^2$

23. $(2c - 5)^2$ **24.** $(6k - 4)^2$

25. $(5x - 3y)^2$ **26.** $(4a - 3b)^2$

27. $(3u - 8v)^2$ **28.** $(4m + 5k)^2$

29. $(9p + 10q)^2$ **30.** $(6c - 7d)^2$

See **31.** $(r + s)(r - s)$ **32.** $(x - y)(x + y)$
Example 27
33. $(u + 2)(u - 2)$ **34.** $(k - 5)(k + 5)$

35. $(3 - a)(3 + a)$ **36.** $(7 - p)(7 + p)$

37. $(2x + 3)(2x - 3)$ **38.** $(4y - 5)(4y + 5)$

39. $(6a - 4b)(6a + 4b)$ **40.** $(3mn - 8)(3mn + 8)$

41. $(9x^2y^2 - 2)(9x^2y^2 + 2)$ **42.** $(10cd - 7f)(10cd + 7f)$

Life science **43.** The population of a bacteria is given by
application
$$P = 5000(3 + t)^2$$

Multiply this expression out.

Business **44.** If \$1000 is compounded for 2 years at interest rate r, the balance B after
application   2 years is given by
$$B = 1000(1 + r)^2$$

Multiply this expression out.

3.7 DIVIDING POLYNOMIALS BY MONOMIALS

Now we wish to divide a polynomial by a monomial. Let us write the polynomial as a sum of monomials $A + B + C + \cdots$. To divide this by another monomial M, we get

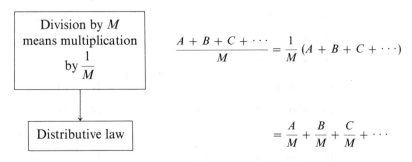

Division by M means multiplication by $\dfrac{1}{M}$

$$\frac{A + B + C + \cdots}{M} = \frac{1}{M}(A + B + C + \cdots)$$

Distributive law

$$= \frac{A}{M} + \frac{B}{M} + \frac{C}{M} + \cdots$$

This becomes our division property: We divide each term in the polynomial by the monomial.

> **PROPERTY 8** For monomials A, B, C, and M,
>
> $$\frac{A + B + C + \cdots}{M} = \frac{A}{M} + \frac{B}{M} + \frac{C}{M} + \cdots$$

EXAMPLE 28   Divide $\dfrac{12a^2b^3 + 8ab^4 - 4a^4b^5}{2ab}$.

SOLUTION   We use the division property above to split this division into three divisions.

| Split into three fractions | $\dfrac{12a^2b^3 + 8ab^4 - 4a^4b^5}{2ab} = \dfrac{12a^2b^3}{2ab} + \dfrac{8ab^4}{2ab} - \dfrac{4a^4b^5}{2ab}$ |

Divide monomials

$$= 6ab^2 + 4b^3 - 2a^3b^4$$

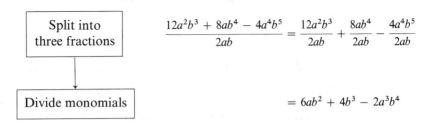

YES	NO
$\dfrac{6x^2 + 12x}{2x} = \dfrac{6x^2}{2x} + \dfrac{12x}{2x}$	~~$\dfrac{2x}{6x^2 + 12x} = \dfrac{2x}{6x^2} + \dfrac{2x}{12x}$~~

We can only "split the numerator."

EXAMPLE 29   Divide $(10x^2y^5 - 15x^3y + 25x^4y^2) \div 5xy^2$.

SOLUTION   We again split this into three fractions and simplify each separately.

Split into three fractions

$$\frac{10x^2y^5 - 15x^3y + 25x^4y^2}{5xy^2} = \frac{10x^2y^5}{5xy^2} - \frac{15x^3y}{5xy^2} + \frac{25x^4y^2}{5xy^2}$$

Divide monomials

$$= 2xy^3 - \frac{3x^2}{y} + 5x^3$$

EXAMPLE 30   Divide $(5x^4 + 10x^3 - 30x + 20) \div 10x^2$.

SOLUTION   We write this as four fractions.

$$\frac{5x^4 + 10x^3 - 30x + 20}{10x^2} = \frac{5x^4}{10x^2} + \frac{10x^3}{10x^2} - \frac{30x}{10x^2} + \frac{20}{10x^2}$$

$$= \frac{x^2}{2} + x - \frac{3}{x} + \frac{2}{x^2}$$

PROBLEM SET 3.7

Divide the following polynomials by $2a$.

See Example 28

1. $8a^2 + 4a$

2. $12a^5 - 6a^3$

3. $4a^3 - 2a$

4. $20a^2 - 4a$

5. $8ab - 6a^4 + 10a^3$

6. $6a^5 - 4a^2b - 12a^5b^2$

7. $18abc - 6a^4 - 10a^3$

8. $2a^2 - 4a^3 + 6a^4 - 8a^5 + 10a^6$

Divide the following polynomials by $5x^2y^3$.

See Example 29

9. $10x^4y^7 - 15x^7y^3$

10. $20x^5y^5 + 30x^3y^4$

11. $5x^3y^5 - 50x^7y^5$

12. $15x^6y^3 - 25x^4y^8$

13. $10x^3y^3 - 15x^2y^4 + 20xy^5$

14. $15x^3y^4 - 30x^2y^5 - 45xy^6$

15. $40x^4y^7 - 30x^2y^3 + 10xy$

16. $25x^3y^3 - 15x^2y^2 + 5xy$

Divide the following polynomials and monomials.

See Example 28

17. $\dfrac{6a^2b - 9ab^4}{3ab}$

18. $\dfrac{4x^2y^2 - 6xy^5}{2xy}$

19. $\dfrac{33u^7v^2 - 12u^2v^7 - 9u^2v^2}{3uv^2}$

20. $\dfrac{25r^3s^2 - 50r^2s + 15r^4s}{5r^2s}$

21. $\dfrac{4m^2n^3k^3 - 6m^4n^3k^3}{2mn^2k^3}$

22. $\dfrac{10a^2b^4c^6 - 25a^9b^7c^5}{5a^2b^4c^5}$

See Examples 29 and 30

23. $(a^7 + a^5 + a^2) \div a^3$

24. $(x^9 - x^5 + x^3) \div x^4$

25. $(33s^2t^2 - 18s^3t + 24s^4t^2) \div 3s^3t^3$

26. $(15u^2v^3 - 10u^3v + 25u^5v^4) \div 5u^2v^2$

27. $(49a^2bc^2 - 35a^3b^2c + 14a^4bc^3) \div 7ab^2c^3$

28. $(42x^2y^3 + 36x^7z^2 - 24y^3z^4) \div 6xyz$

Science
application

29. An object is dropped from a tall building. After t seconds, its distance from the ground is given by

$$s = -16t^2 + 120$$

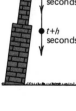

t
seconds

t+h
seconds

(a) Compute the distance S after $(t + h)$ seconds given by

$$S = -16(t + h)^2 + 120$$

(b) Compute the difference of these distances, $S - s$.
(c) Compute the average velocity given by

$$v = \frac{S - s}{h}$$

3.8   **DIVIDING POLYNOMIALS BY POLYNOMIALS**

Dividing two polynomials is like dividing two whole numbers. Consider the following.

EXAMPLE 31   Notice the similarity in the following two divisions.

$$
\begin{array}{r}
23 \\
21{\overline{\smash{\big)}\,487}} \\
42 \\
\hline
67 \\
63 \\
\hline
4
\end{array}
\qquad
\begin{array}{r}
2x + 3 \\
2x + 1{\overline{\smash{\big)}\,4x^2 + 8x + 7}} \\
4x^2 + 2x \\
\hline
6x + 7 \\
6x + 3 \\
\hline
4
\end{array}
$$

As we write $\frac{487}{21} = 23 + \frac{4}{21}$ (or $23\frac{4}{21}$), we write

$$\frac{4x^2 + 8x + 7}{2x + 1} = 2x + 3 + \frac{4}{2x + 1}$$

As a reminder, the key terms in a division problem are presented as follows:

$$
\begin{array}{r}
\text{quotient} \\
\text{divisor} {\overline{\smash{\big)}\,\text{dividend}}} \\
\text{\textasteriskcentered\textasteriskcentered\textasteriskcentered} \\
\hline
\text{\textasteriskcentered\textasteriskcentered\textasteriskcentered} \\
\text{\textasteriskcentered\textasteriskcentered\textasteriskcentered} \\
\hline
\text{remainder}
\end{array}
$$

When we divide polynomials, we use the same procedure over and over throughout the problem: Divide into the leftmost term, multiply by the divisor, subtract, and bring down the next term. We see now how this works.

EXAMPLE 32   Divide $(3x^2 + 11x - 20) \div (x + 5)$.

SOLUTION   We divide like whole numbers. We start with the highest powers.

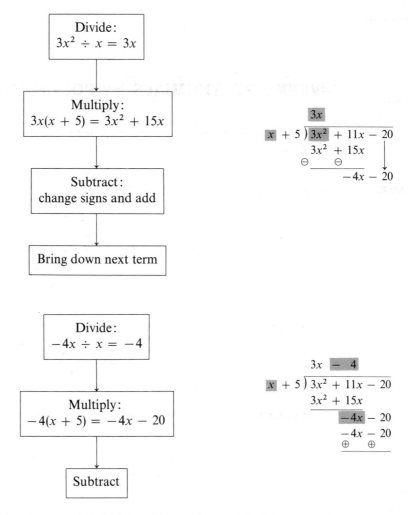

There is no remainder in this problem (or the remainder is 0). Notice that when we subtract, we change all the signs and add.

EXAMPLE 33   Divide $\dfrac{2x^3 - 17x^2 + 29x - 7}{2x - 3}$.

SOLUTION   We start by dividing the highest powers of the divisor and the dividend.

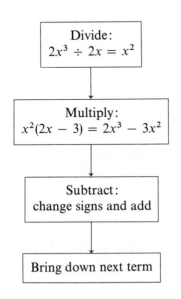

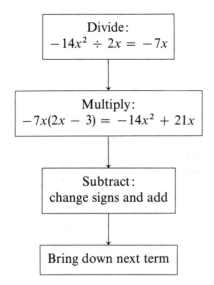

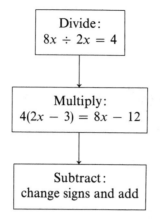

$$\begin{array}{r} x^2 - 7x \quad \boxed{+4} \\ 2x - 3 \overline{\smash{\big)}\, 2x^3 - 17x^2 + 29x - 7} \\ \underline{2x^3 - 3x^2} \\ -14x^2 + 29x \\ \underline{-14x^2 + 21x} \\ \boxed{8x} - 7 \\ 8x - 12 \\ \ominus \qquad \oplus \\ \hline 5 \end{array}$$

We have a remainder of 5. We write the answer as

$$x^2 - 7x + 4 + \frac{5}{2x - 3}$$

EXAMPLE 34   Divide $(6a^3 + 7a^2 - 38a + 11) \div (3a - 4)$.

SOLUTION   We repeatedly divide the highest power of the divisor ($3a$) into the highest remaining power of the dividend.

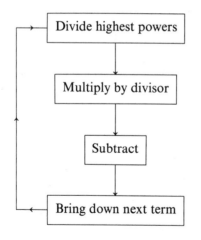

$$\begin{array}{r} 2a^2 + 5a - 6 \\ 3a - 4 \overline{\smash{\big)}\, 6a^3 + 7a^2 - 38a + 11} \\ \underline{6a^3 - 8a^2} \\ \ominus \quad \oplus \\ 15a^2 - 38a \\ 15a^2 - 20a \\ \ominus \qquad \oplus \\ -18a + 11 \\ -18a + 24 \\ \oplus \qquad \ominus \\ \hline -13 \end{array}$$

Thus, the answer is $2a^2 + 5a - 6 + \dfrac{-13}{3a - 4}$.

EXAMPLE 35   Divide $\dfrac{x^5 - 1}{x^2 + 1}$.

SOLUTION Here we have some missing powers of x. It is best to fill in their places with zeros. For example, $x^2 + 1 = x^2 + 0x + 1$. (The zero holds a place as it does in the whole number 101.)

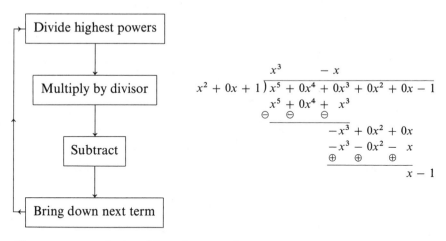

We cannot go any further. Thus, the answer is

$$x^3 - x + \frac{x - 1}{x^2 + 1}$$

PROBLEM SET 3.8

For review, divide the following whole numbers and leave the answers as mixed numbers.

See Example 31

1. $\dfrac{523}{29}$

2. $\dfrac{703}{34}$

3. $\dfrac{1217}{53}$

4. $\dfrac{2135}{79}$

5. $\dfrac{13,062}{92}$

6. $\dfrac{31,548}{123}$

Divide the following polynomials.

See Example 32

7. $\dfrac{x^2 + 12x + 35}{x + 5}$

8. $\dfrac{a^2 - 2a - 24}{a - 6}$

9. $\dfrac{x^2 - 10x + 16}{x - 2}$

10. $\dfrac{r^2 + 2r - 63}{r + 9}$

11. $\dfrac{2x^2 + 11x + 12}{x + 4}$

12. $\dfrac{3k^2 + 16k - 35}{k + 7}$

13. $\dfrac{6t^2 - 19t - 7}{3t + 1}$

14. $\dfrac{12x^2 - 23x + 10}{4x - 5}$

15. $\dfrac{10m^3 + 19m^2 - 13m + 5}{2m + 5}$

16. $\dfrac{8z^3 - 14z^2 - 13z + 4}{4z - 1}$

17. $\dfrac{24u^3 - 11u^2 + 13u - 4}{3u - 1}$

18. $\dfrac{14w^3 + 48w^2 + 14w - 12}{2w + 6}$

*See
Examples
33 and 34*

19. $\dfrac{x^2 + 5x - 7}{x - 4}$

20. $\dfrac{a^2 - 6a + 10}{a + 3}$

21. $\dfrac{w^2 - 6w - 5}{w + 8}$

22. $\dfrac{2r^2 - 3r + 6}{r - 5}$

23. $\dfrac{3k^2 - 6k + 7}{k - 5}$

24. $\dfrac{2t^2 - 7t - 3}{t + 4}$

25. $\dfrac{4x^3 - 4x^2 - 23x + 15}{2x - 5}$

26. $\dfrac{15y^3 + 4y^2 + 7y + 4}{5y + 3}$

27. $\dfrac{8u^3 - 32u^2 + 8u + 17}{2u - 7}$

28. $\dfrac{10r^3 - 26r^2 - 22r + 14}{2r - 6}$

29. $\dfrac{20m^3 - 21m^2 + 19m + 1}{5m - 4}$

30. $\dfrac{14x^3 + 27x^2 - 24x + 2}{2x + 5}$

*See
Example 35*

31. $\dfrac{x^3 - 1}{x - 1}$

32. $\dfrac{t^3 + 1}{t + 1}$

33. $\dfrac{x^4 + 1}{x + 1}$

34. $\dfrac{r^5 - 1}{r + 1}$

35. $\dfrac{u^6 + u^3 + 2}{u^3 - 1}$

36. $\dfrac{m^7 - m^4 - 1}{m^3 - 1}$

CHAPTER 3 SUMMARY

Important Words and Phrases

base (3.1)
binomial (3.4)
coefficient (3.3)

degree (3.4)
dividend (3.8)
divisor (3.8)

exponent (3.1)

power (3.1)

exponent form (3.1)

quotient (3.8)

FOIL (3.6)

remainder (3.8)

like terms (3.3)

square of a binomial (3.6)

monomial (3.3)

sum and difference property (3.6)

negative exponent (3.2)

term (3.3)

opposite of a polynomial (3.4)

trinomial (3.4)

polynomial (3.4)

zero exponent (3.2)

Important Properties and Definitions

For any real numbers a, b, c, and d and positive integers m, n, k,

$$a^n = a \cdot a \cdot \ldots \cdot a \quad (n\text{-factors})$$

$$a^m a^n = a^{m+n}$$

$$(a^m)^n = a^{m \cdot n}$$

$$(ab)^k = a^k b^k$$

$$\frac{a^m}{a^n} = \begin{cases} a^{m-n} & \text{if } m > n \\ 1 & \text{if } m = n \quad (a \neq 0) \\ \dfrac{1}{a^{n-m}} & \text{if } n > m \end{cases}$$

$$\frac{a^m}{a^n} = a^{m-n} \quad (a \neq 0)$$

$$\left(\frac{a}{b}\right)^k = \frac{a^k}{b^k} \quad (b \neq 0)$$

$$a^{-n} = \frac{1}{a^n} \quad (a \neq 0)$$

$$a^0 = 1 \quad (a \neq 0)$$

$$(a + b)(c + d) = ac + ad + bc + bd$$

$$(a + b)^2 = a^2 + 2ab + b^2$$

$$(a - b)^2 = a^2 - 2ab + b^2$$

$$(a + b)(a - b) = a^2 - b^2$$

$$\frac{a + b + c + \cdots}{m} = \frac{a}{m} + \frac{b}{m} + \frac{c}{m} + \cdots \quad (m \neq 0)$$

Important Procedures

To add like monomials, add the coefficients.

To multiply (divide) monomials:
1. Multiply (divide) the coefficients.
2. Add (subtract) exponents of the same base.

To add polynomials:
1. Group (or line up) like terms.
2. Add the coefficients.

To subtract polynomials:
1. Change *all* the signs in the polynomial to be subtracted (find its opposite).
2. Add the resulting polynomials.

To multiply polynomials:
1. Multiply all the terms of one polynomial by all the terms of the other polynomial.
2. Add the like terms.

To divide polynomials:
1. Divide highest powers.
2. Multiply by the divisor.
3. Subtract. (Change signs and add.)
4. Bring down the next term.
5. Repeat this process until there is a remainder that cannot be divided. Write the answer as

$$\text{quotient} + \frac{\text{remainder}}{\text{divisor}}$$

CHAPTER 3 REVIEW EXERCISES

Write the following in exponent form.

1. $t \cdot t \cdot t \cdot t \cdot t$ **2.** $8 \cdot 8 \cdot 8$

Write each of the following as a single number.

3. 2^5 **4.** $(-10)^2$

Simplify each of the following as much as possible.

5. $x^2 x^4 x^5$

6. $(a^4)^5$

7. $(3uv^2)^3$

8. $\dfrac{m^5}{m^3}$

9. $\dfrac{n^4}{n^9}$

10. $\left(\dfrac{2}{k}\right)^4$

11. $\left(\dfrac{4a^2 b}{c^3}\right)^2$

12. 2^{-5}

13. 4^0

Give the coefficients of the following monomials.

14. $6abc$

15. $-x^2 y$

Give the degree and special name for the following polynomials.

16. $t^2 - 5t + 6$

17. $6x^4 + 12$

Perform the indicated operations.

18. $7pq - 9pq - 3pq$

19. $(4x^2 y)(3xy^5)(-2x^2 y^2)$

20. $\dfrac{40a^2 b^3 c^8}{-10a^3 b^3 c^3}$

21. Add $\quad\begin{array}{r} 4t^2 - 5t + 1 \\ 3t^2 + 8t - 4 \\ 8t^3 - 2t^2 - 6t + 8 \\ \hline \end{array}$

22. $(4u^4 - 3u^3 - 2u) + (5u^4 - 7u^2 + 5) + (6u^3 - 7u + 2)$

23. $(6x^3 - 4x^2 + 10x - 1) - (7x^3 + 8x^2 + 7x - 4)$

24. $(3m^2 - 8) + (7m^2 + 7) - (5m^2 - 3)$

25. $6a^2(5a^4 - 3a^2 - a)$

26. $(3u - 5)(8u + 3)$

27. $(x + 4)(2x^2 + 5x + 7)$

28. $(x + 1)(x + 4)(x - 7)$

29. $(t^2 + t - 1)(2t^3 - 5t^2 + t + 3)$

30. $(6r - 3)(4r + 5)$

31. $(2x - 5)^2$

32. $(3a + 2)^2$

33. $(7p - 5q)(7p + 5q)$

34. $\dfrac{6x^2 y^2 - 12x^3 y^4 + 18x^4 y^5}{2xy^2}$

35. $\dfrac{10a^2 b^3 - 15ab^4 + 20a^7 b}{5a^2 b^2}$

36. $\dfrac{x^2 + 3x - 40}{x - 5}$

37. $\dfrac{6x^3 - 19x^2 + 27x - 20}{3x - 5}$

38. $\dfrac{6u^3 + 25u^2 + 21u - 4}{2u + 5}$

39. $\dfrac{a^5 - 1}{a - 1}$

4
FACTORING

4.1   FACTORING NATURAL NUMBERS

Factoring is the opposite of multiplying. When multiplying, we have two factors, and we want to find the product:

Multiplication:
$$5 \cdot 7 = \boxed{?}$$
$$(x + 7)(2x - 3) = \boxed{\quad ? \quad}$$

When factoring, we have the product, and we want to find the factors that produced that product:

Factoring:
$$21 = \boxed{?} \cdot \boxed{?}$$
$$x^2 - 3x - 10 = \boxed{\quad ? \quad} \cdot \boxed{\quad ? \quad}$$

Factoring is the process that we use to split a number or polynomial into its factors. In this section we review the factoring of natural numbers.

A natural number, two or more, is said to be **prime** if the only natural numbers that divide it evenly are 1 and itself. For example, 7 is a prime since only 1 and 7 divide 7; however, 12 is *not* prime since 12 can be divided evenly by 1, 2, 3, 4, 6, and 12. The first few primes are

$$2, \quad 3, \quad 5, \quad 7, \quad 11, \quad 13, \quad 17, \quad \ldots$$

A natural number, greater than 2, that is not a prime is called a **composite.**

> Every composite number can be written as a product of primes.

EXAMPLE 1 $15 = 3 \cdot 5.$

EXAMPLE 2   Factor 36 into a product of primes.

SOLUTION   We can use a "factor tree" to help factor 36. We continue factoring until we are down only to primes.

Thus, $36 = 2 \cdot 2 \cdot 3 \cdot 3$. Notice that both trees yield the same primes.

EXAMPLE 3   Factor 84 into a product of primes.

SOLUTION   Here we factor out all 2's and then 3's.

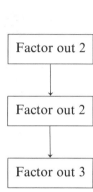

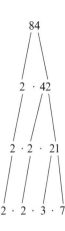

Thus, $84 = 2 \cdot 2 \cdot 3 \cdot 7$.

EXAMPLE 4   Factor 71 into a product of primes.

SOLUTION   We look for factors of 71, but we find only 1 and 71. Thus, 71 is a prime.

PROBLEM SET 4.1

For the following natural numbers, identify the primes (see Example 4) and write the composites as a product of primes.

See Examples 1, 2, 3, and 4

1. 21	**2.** 12	**3.** 9
4. 35	**5.** 42	**6.** 14
7. 25	**8.** 26	**9.** 27

10. 70	**11.** 100	**12.** 101
13. 105	**14.** 86	**15.** 48
16. 65	**17.** 200	**18.** 160
19. 270	**20.** 243	**21.** 157
22. 143	**23.** 147	**24.** 107
25. 85	**26.** 132	**27.** 175
28. 89	**29.** 99	**30.** 102

4.2   GREATEST COMMON FACTOR

For a set of integers, the **greatest common factor (GCF)** is the largest integer that will divide evenly into all of them. For instance,

$$\text{GCF of } \{12, 18, 30\} = 6$$

since 6 divides evenly into 12, 18, and 30, and no larger number will.

Divisors of 12:   1, 2, 3, 4, ⑥, 12
Divisors of 18:   1, 2, 3, ⑥, 9, 18
Divisors of 30:   1, 2, 3, 5, ⑥, 10, 15, 30

EXAMPLE 5   The following are examples of the GCF of sets of integers.

(a) GCF of $\{10, 15, 25\} = 5$
(b) GCF of $\{20, 50, 100, 200\} = 10$
(c) GCF of $\{40, 16\} = 8$
(d) GCF of $\{9, 16, 25\} = 1$

In example **(d)** notice that the GCF was 1 since there is no number larger than 1 that evenly divides all three.

With algebraic monomials, the greatest common factor (GCF) has a similar definition: It is the monomial with the largest coefficient and highest exponents that will evenly divide all of the given monomials. For instance, the GCF of $8x^2$ and $10x$ is $2x$, since 2 is the GCF of 8 and 10 and x is the highest power of x that divides x and x^2.

EXAMPLE 6   The following are examples of the GCF of monomials.

(a) GCF of $\{2x^2, 4x\} = 2x$

(b) GCF of $\{15a^3, 10a^2, 25a^4\} = 5a^2$

(c) GCF of $\{6x^2y^4, 15x^3y\} = 3x^2y$

(d) GCF of $\{4r^2s^3t^4, 6r^6st^3, 12r^5s^4t^2\} = 2r^2st^2$

(e) GCF of $\{ab, ac, bc\} = 1$

Notice that in every case the exponent of each variable in the GCF is the smallest of the exponents that the variables were raised to.

We use the GCF idea along with the distributive law to help us factor polynomials.

> Any polynomial factors into:
>
> **1.** The GCF of all its terms times
> **2.** The polynomial divided by the GCF.

EXAMPLE 7   Factor $5x + 10y$.

SOLUTION   The GCF of the two terms is 5. Using the distributive law, we can factor this as follows.

$$\boxed{\text{Divide out GCF, 5}} \qquad\qquad 5x + 10y = 5(x + 2y)$$

EXAMPLE 8   Factor $4a^3 + 8a^2 - 16a$.

SOLUTION   Here, the GCF is $4a$. We divide this out of all the terms.

$$\boxed{\text{Divide out GCF, } 4a} \qquad\qquad 4a^3 + 8a^2 - 16a = 4a(a^2 + 2a - 4)$$

EXAMPLE 9   Factor $10r^2s^3 - 15r^5s - 25r^3s^4$.

SOLUTION   The GCF of the three terms is $5r^2s$, which we divide from each term of the polynomial.

$$\boxed{\begin{array}{c}\text{Divide out GCF,}\\ 5r^2s\end{array}} \qquad 10r^2s^3 - 15r^5s - 25r^3s^4 = 5r^2s(2s^2 - 3r^3 - 5rs^3)$$

EXAMPLE 10   Factor $9a^2bc^3 - 6ab^3c + 12a^3b^2c$.

SOLUTiON   The GCF here is $3abc$.

| Divide out GCF, $3abc$ | $9a^2bc^3 - 6ab^3c + 12a^3b^2c = 3abc(3ac^2 - 2b^2 + 4a^2b)$ |

YES	NO
$6a^3 + 2a = 2a(3a^2 + 1)$	~~$6a^3 + 2a = 2a(3a^2)$~~

We must factor the GCF out of *all* the terms.

PROBLEM SET 4.2

Find the greatest common factor (GCF) for each of the following sets of numbers and monomials. [*Be careful;* some may have GCF = 1—see Examples 5(**d**) and 6(**e**).]

*See
Example 5*
1. $\{8, 12\}$ **2.** $\{10, 15\}$

3. $\{6, 12, 18\}$ **4.** $\{20, 40, 80\}$

5. $\{15, 30, 40\}$ **6.** $\{27, 12, 8\}$

*See
Example 6*
7. $\{3x, 6x^2\}$ **8.** $\{7a^2, 14a\}$

9. $\{ab^2, a^2b\}$ **10.** $\{4u^3v, 6u^2v^4\}$

11. $\{m^3n^4, m^2n^5, mn^6\}$ **12.** $\{5r^2t, 6r^3s^2, 7st^4\}$

13. $\{18p^2q^5r, 9p^4q^6, 27p^3q^4\}$ **14.** $\{20a^2b^3c, 10ab^4c^3, 25a^4b^2c^5\}$

Factor the following polynomials by factoring out the GCF.

*See
Example 7*
15. $2x + 6y$ **16.** $10a - 15b$

17. $16r - 8s$ **18.** $20u + 10v$

19. $25x + 35y$ **20.** $40m - 48n$

*See
Example 8*
21. $8x - 4x^2$ **22.** $2a - 14a^3$

23. $6t^3 - 9t^2$ **24.** $12u^4 + 18u^6$

25. $2k^2 - 4k^3 + 8k^4$ **26.** $3m^3 + 6m^5 - 9m^7$

27. $10u^3 - 12u^7 + 24u^{16}$ **28.** $21p^5 - 18p^4 + 15p^3$

See   **29.** $a^4b^3 + a^3b^4$ **30.** $u^7v^2 - u^5v^3$

Example 9

31. $2x^2y^4 - 6x^3y$ **32.** $9r^2s - 6rs^4$

33. $10u^4v - 15u^2v^2 + 20u^3v^4$ **34.** $16m^2n^3 - 24m^4n + 32m^3n^5$

35. $18x^7y^2 - 24x^3y^5 + 30x^4y$ **36.** $25a^3b^7 - 35a^2b^6 - 45a^4b^{10}$

See   **37.** $u^2v^3w^4 + u^5vw^2$ **38.** $a^2b^3c^5 - a^5bc^7$

Example 10

39. $8k^4m^3n^{10} - 12k^2mn^7$ **40.** $16b^4c^9d^7 - 20c^6d^8e^{10}$

41. $4xy^2z - 8x^2yz - 12xyz^2$ **42.** $15u^2v^2w - 20uv^3w - 25u^2vw^3$

Business application   **43.** If P dollars is invested at a rate of interest r, after 1 year this is worth $P + Pr$. Factor this expression.

Health application   **44.** The rate at which a certain epidemic spreads through a town is given $100\,y - 0.01y^2$. Factor this expression.

4.3   FACTORING TRINOMIALS

Factoring is the reverse of multiplication. Since the product of two binomials is usually a trinomial, it is reasonable to try to factor a trinomial into two binomials. In this section we only look at trinomials whose square term has coefficient 1, such as $x^2 - 4x - 12$.

Let us now review how the FOIL method worked when multiplying binomials.

$$(x + 5)(x + 6) = x^2 \quad\;\; + 11x \;\; + 30$$
$$(x - 9)(x + 2) = x^2 \quad\;\; - 7x \;\; - 18$$
$$(x - 8)(x - 7) = x^2 \quad\;\; - 15x \;\; + 56$$
$$(x + a)(x + b) = x^2 + (a + b)x + ab$$
$$\qquad\qquad\qquad\qquad\uparrow\qquad\;\;\uparrow$$
$$\qquad\qquad\qquad\quad\text{sum}\quad\text{product}$$

Notice the pattern. The coefficient of the middle term is always the *sum*, and the last term always the *product* of the second terms of the binomials. This suggests a general procedure.

> To factor a trinomial of the type $x^2 + Mx + N$ (M and N integers):
>
> **1.** Find all pairs of factors for the last term, N.
>
> **2.** Choose the pair of factors (if it exists), a and b, whose sum is the middle coefficient, M.
>
> **3.** The factors are $(x + a)(x + b)$.
>
> **4.** Check the factoring by multiplying using the FOIL method.

EXAMPLE 11 Factor $x^2 + 7x + 10$.

SOLUTION We look for the factors of 10 that add up to 7.

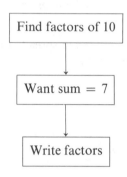

Factors of 10	Sum
1, 10	11
2, 5 ✓	7 ✓
−1, −10	−11
−2, −5	−7

$$x^2 + 7x + 10 = (x + 2)(x + 5)$$

We can check this by multiplying $(x + 2)(x + 5) = x^2 + 7x + 10$.

EXAMPLE 12 Factor $a^2 + 2a - 24$.

SOLUTION Here we factor -24 and look for a pair of factors that adds up to 2.

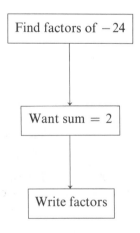

Factors of −24	Sum
1, −24	−23
2, −12	−10
3, −8	−5
4, −6	−2
6, −4 ✓	2 ✓
8, −3	5
12, −2	10
24, −1	23

$$a^2 + 2a - 24 = (a + 6)(a - 4)$$

When working a problem such as this, a student would not usually write out all the combinations. Rather, he or she would go through them mentally and stop at the right pair. We can check this by multiplying $(a + 6)(a - 4) = a^2 + 2a - 24$.

EXAMPLE 13   Factor $t^2 - 9t + 18$.

SOLUTION   We look at all the factors of 18 and try to find the pair whose sum is -9.

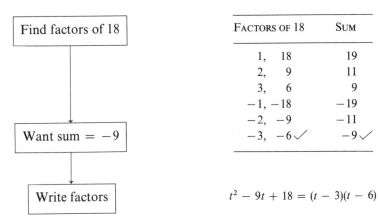

FACTORS OF 18	SUM
1, 18	19
2, 9	11
3, 6	9
$-1, -18$	-19
$-2, -9$	-11
$-3, -6$ ✓	-9 ✓

$$t^2 - 9t + 18 = (t - 3)(t - 6)$$

We check this by multiplying: $(t - 3)(t - 6) = t^2 - 9t + 18$.

EXAMPLE 14   Factor $x^2 + 5x - 8$.

SOLUTION   We begin by factoring -8 and looking for the factors whose sum is 5.

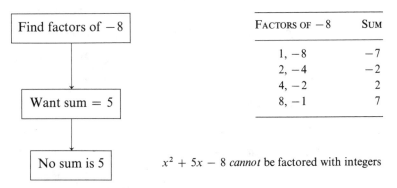

FACTORS OF -8	SUM
1, -8	-7
2, -4	-2
4, -2	2
8, -1	7

$x^2 + 5x - 8$ *cannot* be factored with integers

Just as a prime number such as 7 cannot be factored (except as $1 \cdot 7$), there are polynomials, such as $x^2 + 5x - 8$, that cannot be factored with integers. This is the case here.

EXAMPLE 15   Factor $2x^3 + 2x^2 - 40x$.

SOLUTION   We first factor out the GCF of $2x$. Then we look for the binomial factors of the remaining trinomial.

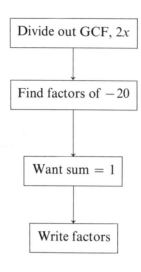

$$2x^3 + 2x^2 - 40x = 2x(x^2 + x - 20)$$

FACTORS OF -20	SUM
$1, -20$	-19
$2, -10$	-8
$4, -5$	-1
$5, -4 \checkmark$	$1 \checkmark$
$10, -2$	8
$20, -1$	19

$$2x^3 + 2x^2 - 40x = 2x(x + 5)(x - 4)$$

PROBLEM SET 4.3

Factor the following trinomials. (*Be careful;* some may not factor with integers—see Example 14.)

See
Examples
11, 12, and 13

1. $x^2 + 4x + 3$
2. $u^2 + 6u + 8$
3. $a^2 + 8a + 12$
4. $m^2 - 2m - 3$
5. $r^2 + 8r + 15$
6. $k^2 - 6k - 16$
7. $p^2 + 3p - 18$
8. $n^2 + 6n - 16$
9. $t^2 - 7t + 10$
10. $q^2 + 3q - 54$
11. $w^2 - 10w + 15$
12. $z^2 - z - 12$
13. $b^2 - 8b + 12$
14. $y^2 - 16y + 48$
15. $x^2 + 8x + 7$
16. $a^2 - 11a + 18$
17. $d^2 + 5d - 24$
18. $u^2 + 16u + 28$
19. $y^2 + y - 20$
20. $t^2 - 12t - 45$
21. $s^2 + 2s - 3$
22. $v^2 - 6v - 27$
23. $r^2 + 2r - 15$
24. $m^2 - 7m + 6$
25. $k^2 - 10k + 24$
26. $x^2 - 10x + 16$

See
Example 15

27. $2x^2 - 8x - 10$ **28.** $4a^2 + 8a - 192$

29. $10r^2 + 60r + 80$ **30.** $5k^2 - 25k - 180$

31. $-2x^2 + 8x + 42$ **32.** $3u^2 + 3u - 126$

33. $t^3 + 4t^2 - 45t$ **34.** $m^4 - m^3 - 72m^2$

35. $2r^3 + 10r^2 - 12r$ **36.** $3p^4 - 33p^3 - 36p^2$

37. $6z^5 + 6z^4 - 12z^3$ **38.** $-2u^3 - 18u^2 + 44u$

39. $4y^6 - 20y^5 - 24y^4$ **40.** $5x^4 - 15x^3 - 140x^2$

*Science
application*
41. The height h of an object shot into the air from below the ground is given by $h = -16t^2 + 80t - 96$. Factor the polynomial on the right completely.

*Business
application*
42. The *marginal cost MC* to produce x items is given by

$$MC = 3x^2 - 12x + 12$$

Factor the polynomial on the right completely.

4.4   MORE TRINOMIAL FACTORING

In this section we continue factoring trinomials. Now, however, the trinomials will have a coefficient for the square term that is *not* 1, such as $6x^2 - 13x - 5$. Let us again review the FOIL method for multiplying binomials.

$$(2x + 1)(3x + 2) = 6x^2 + \;\;7x + \;\;2$$
$$(4x - 3)(2x + 5) = 8x^2 + 14x - 15$$
$$(4x + 3)(2x - 5) = 8x^2 - 14x - 15$$

$$F = product\ of\ firsts \qquad L = product\ of\ lasts$$
$$O + I = sum\ of\ outer\text{-}product\ and\ inner\text{-}product$$

The first and third terms of the product (F and L) are the easiest to work with, so we start with them. Also, look at the last two examples. Notice that the signs within the binomials have been reversed, but this only changes the sign of the middle term. Let us now give the procedure for factoring these trinomials. We follow this with examples.

> To factor the general trinomial $ax^2 + bx + c$:
>
> **1.** Factor the first term, ax^2.
> **2.** Factor the last term, c.
> **3.** Look at all the binomial combinations of the factors of ax^2 and the factors of c.
> **4.** Find the combination (if it exists) that gives the middle term, bx.
> **5.** If you get the middle term to be $-bx$ (right absolute value, wrong sign), then reverse *both* signs of the factors of c.

EXAMPLE 16   Factor $2x^2 + 11x + 15$.

SOLUTION   We begin by factoring the first and last terms. We then look for a combination that will give the middle term, $11x$.

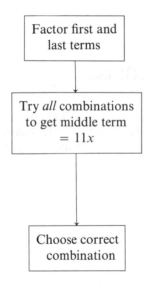

COMBINATIONS	MIDDLE TERM
$(2x + 1)(x + 15)$	$30x + x = 31x$
$(2x + 15)(x + 1)$	$2x + 15x = 17x$
$(2x + 3)(x + 5)$	$10x + 3x = 13x$
$(2x + 5)(x + 3)\checkmark$	$6x + 5x = 11x\checkmark$

$$2x^2 + 11x + 15 = (2x + 5)(x + 3)$$

We check this by multiplying: $(2x + 5)(x + 3) = 2x^2 + 11x + 15$.

EXAMPLE 17   Factor $3x^2 + 5x - 12$.

SOLUTION   We factor the first and last terms and look for a combination that gives a middle term of $5x$.

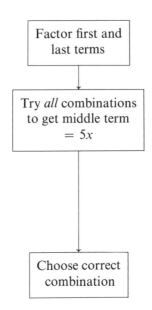

$$3x^2 \qquad\qquad -12$$
$$(3x\ \ )(x\ \ ) \longleftrightarrow \begin{array}{l}(\ \ +1)(\ \ -12)\\(\ \ +2)(\ \ \ -6)\\(\ \ +3)(\ \ \ -4)\end{array}$$

COMBINATIONS	MIDDLE TERM
$(3x + 1)(x - 12)$	$-36x + x = -35x$
$(3x - 12)(x + 1)$	$3x - 12x = -9x$
$(3x + 2)(x - 6)$	$-18x + 2x = -16x$
$(3x - 6)(x + 2)$	$6x - 6x = 0$
$(3x + 3)(x - 4)$	$-12x + 3x = -9x$
$(3x - 4)(x + 3)\checkmark$	$9x - 4x = 5x\checkmark$

Factor first and last terms

Try *all* combinations to get middle term $= 5x$

Choose correct combination

$$3x^2 + 5x - 12 = (3x - 4)(x + 3)$$

EXAMPLE 18   Factor $6a^2 + 17a - 14$.

SOLUTION   We first find the factors of $6a^2$ and -14. Then we look for the combination that will give a middle term of $17a$.

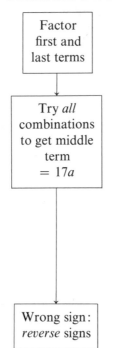

$$6a^2 \qquad\qquad -14$$
$$\begin{array}{l}(6a\ \ )(a\ \ )\\(2a\ \ )(3a\ \ )\end{array} \longleftrightarrow \begin{array}{l}(\ \ +1)(\ \ -14)\\(\ \ +2)(\ \ \ -7)\end{array}$$

Factor first and last terms

Try *all* combinations to get middle term $= 17a$

COMBINATIONS	MIDDLE TERM
$(6a + 1)(a - 14)$	$-84a + a = -83a$
$(6a - 14)(a + 1)$	$6a - 14a = -8a$
$(6a + 2)(a - 7)$	$-42a + 2a = -40a$
$(6a - 7)(a + 2)$	$12a - 7a = 5a$
$(2a + 1)(3a - 14)$	$-28a + 3a = -25a$
$(2a - 14)(3a + 1)$	$2a - 42a = -40a$
$(2a + 2)(3a - 7)$	$-14a + 6a = -8a$
$(2a - 7)(3a + 2)$	$4a - 21a = -17a$ (almost)$\checkmark$

Wrong sign: *reverse* signs

$$6a^2 + 17a - 14 = (2a + 7)(3a - 2)$$

Notice that we found a middle term of $-17a$ (wrong sign), so we reversed the signs in *both* factors of -14. Thus, we changed $(2a - 7)(3a + 2)$ to $(2a + 7)(3a - 2)$. We can check by multiplying $(2a + 7)(3a - 2) = 6a^2 + 17a - 14$.

EXAMPLE 19   Factor $2x^2 + 9x + 21$.

SOLUTION   We begin by factoring $2x^2$ and 21. We then look for a combination that will give a middle term of $9x$.

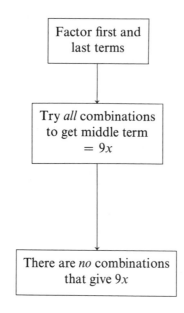

COMBINATIONS	MIDDLE TERM
$(2x + 1)(x + 21)$	$42x + x = 43x$
$(2x + 21)(x + 1)$	$2x + 21x = 23x$
$(2x + 3)(x + 7)$	$14x + 3x = 17x$
$(2x + 7)(x + 3)$	$6x + 7x = 13x$

$2x^2 + 9x + 21$ cannot be factored with integers

Thus, the polynomial $2x^2 + 9x + 21$ cannot be factored with integers.

EXAMPLE 20   Factor $40x^4 - 50x^3 - 90x^2$.

SOLUTION   Before we factor this as a trinomial, let us first factor out the GCF, $10x^2$. Then we factor the remaining polynomial.

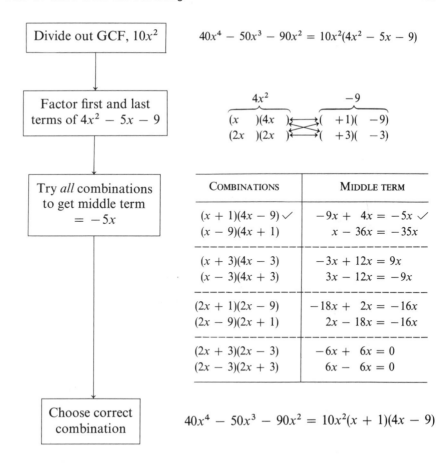

| Divide out GCF, $10x^2$ | $40x^4 - 50x^3 - 90x^2 = 10x^2(4x^2 - 5x - 9)$ |

Factor first and last terms of $4x^2 - 5x - 9$

$$
\begin{array}{cc}
\overbrace{4x^2} & \overbrace{-9} \\
(x\ )(4x\ ) & (\ +1)(\ -9) \\
(2x\ )(2x\ ) & (\ +3)(\ -3)
\end{array}
$$

Try *all* combinations to get middle term $= -5x$

COMBINATIONS	MIDDLE TERM
$(x + 1)(4x - 9)$ ✓	$-9x + 4x = -5x$ ✓
$(x - 9)(4x + 1)$	$x - 36x = -35x$
$(x + 3)(4x - 3)$	$-3x + 12x = 9x$
$(x - 3)(4x + 3)$	$3x - 12x = -9x$
$(2x + 1)(2x - 9)$	$-18x + 2x = -16x$
$(2x - 9)(2x + 1)$	$2x - 18x = -16x$
$(2x + 3)(2x - 3)$	$-6x + 6x = 0$
$(2x - 3)(2x + 3)$	$6x - 6x = 0$

Choose correct combination

$$40x^4 - 50x^3 - 90x^2 = 10x^2(x + 1)(4x - 9)$$

PROBLEM SET 4.4

Factor the following trinomials. (*Be careful;* some may not factor with integers—see Example 19.)

See Examples 16, 17, and 18

1. $2x^2 + 5x + 3$

2. $2a^2 + 7a + 5$

3. $2u^2 + 11u + 5$

4. $3r^2 + 16r + 5$

5. $8y^2 - 2y - 3$

6. $8z^2 - 18z - 35$

7. $3x^2 - 19x - 14$

8. $10m^2 + 19m + 6$

9. $5k^2 + 28k - 12$

10. $10t^2 + 43t + 28$

11. $10y^2 - 13y - 3$

12. $16w^2 + 38w - 5$

13. $8p^2 - 9p + 15$	**14.** $6u^2 - 37u + 45$
15. $5a^2 - 31a - 28$	**16.** $2v^2 + 25v + 50$
17. $6q^2 - 19q + 15$	**18.** $12x^2 + 40x - 7$
19. $6z^2 + 23z + 15$	**20.** $6n^2 - 31n + 35$
21. $8p^2 - 26p + 15$	**22.** $15m^2 - 29m - 14$
23. $10x^2 - 53x + 15$	**24.** $6z^2 - 13z - 10$
25. $8k^2 - 53k - 21$	**26.** $20q^2 - 47q + 24$
27. $30r^2 + 43r + 15$	**28.** $12u^2 + 40u - 63$
29. $35t^2 + 11t - 6$	**30.** $20y^2 - 103y + 15$

See Example 20

31. $12a^3 + 14a^2 + 4a$	**32.** $12x^4 + 24x^3 - 63x^2$
33. $120t^4 - 50t^3 - 20t^2$	**34.** $12r^5 + 70r^4 + 72r^3$
35. $30u^3 + 9u^2 - 12u$	**36.** $8m^2 - 28m - 60$
37. $24k^4 - 34k^3 + 12k^2$	**38.** $14x^9 - 65x^8 + 9x^7$
39. $5y^4 + 42y^3 + 16y^2$	**40.** $-30b^2 + 8b + 64$
41. $-21a^5 + 19a^4 + 12a^3$	**42.** $6t^9 + 29t^8 + 9t^7$
43. $16x^2 - 44x + 30$	**44.** $10m^5 - 11m^4 - 6m^3$

4.5   SPECIAL FACTORINGS

In Section 3.6 we saw the following special binomial products:

$$(a + b)(a - b) = a^2 - b^2$$
$$(a + b)^2 = a^2 + 2ab + b^2$$
$$(a - b)^2 = a^2 - 2ab + b^2$$

In this section we reverse these. We look for products in the forms on the right above and use the forms on the left as factorings. Let us start with the **difference-of-squares** property.

> **PROPERTY 1**   For any real numbers a and b,
>
> $$a^2 - b^2 = (a + b)(a - b)$$

EXAMPLE 21   The following are all differences of squares and use Property 1 to factor.

(a) $x^2 - 25 = (x + 5)(x - 5)$

(b) $a^2 - 49 = (a + 7)(a - 7)$

(c) $k^2 - 100 = (k - 10)(k + 10)$

Notice that it does not matter which of the factors we write first.

EXAMPLE 22   Factor $25m^2 - 81$.

SOLUTION   This is also a difference of squares.

$$\boxed{\text{Rewrite}} \qquad\qquad 25m^2 - 81 = (5m)^2 - 9^2$$

$$\boxed{a^2 - b^2 = (a + b)(a - b)} \qquad\qquad = (5m + 9)(5m - 9)$$

EXAMPLE 23   The following also use the difference-of-squares property.

(a) $4p^2 - 25 = (2p + 5)(2p - 5)$

(b) $100r^2 - 49s^2 = (10r - 7s)(10r + 7s)$

(c) $64t^2 - 9w^2 = (8t + 3w)(8t - 3w)$

YES	NO
$a^2 - 36 = (a - 6)(a + 6)$	~~$a^2 - 36 = (a - 6)(a - 6)$~~
$a^2 + 9$ cannot be factored with real numbers	~~$a^2 + 9 = (a + 3)(a + 3)$~~

Why can $a^2 + 9$ not be factored? The different factors of 9 give the following possibilities:

$$(a + 1)(a + 9) = a^2 + 10a + 9 \qquad \text{(wrong)}$$

$$(a + 3)(a + 3) = a^2 + 6a + 9 \qquad \text{(wrong)}$$

$$(a - 1)(a - 9) = a^2 - 10a + 9 \qquad \text{(wrong)}$$

$$(a - 3)(a - 3) = a^2 - 6a + 9 \qquad \text{(wrong)}$$

None of these work, so there is no factoring for $a^2 + 9$.

EXAMPLE 24   Factor $p^4 - 16$ completely.

SOLUTION   This is a difference of two squares. But once we factor, we have to factor *again*.

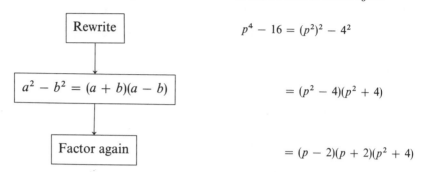

$$p^4 - 16 = (p^2)^2 - 4^2$$

$$= (p^2 - 4)(p^2 + 4)$$

$$= (p - 2)(p + 2)(p^2 + 4)$$

The other special factoring that we use is for the **perfect-square trinomial,** which is the square of a binomial. For instance, $x^2 + 10x + 25 = (x + 5)^2$ is a perfect-square trinomial. We have the following property.

PROPERTY 2   For any real numbers a and b,

$$a^2 + 2ab + b^2 = (a + b)^2$$
$$a^2 - 2ab + b^2 = (a - b)^2$$

To factor a perfect-square trinomial:

1. The first term must be a perfect square, a^2.
2. The last term must be a perfect square, b^2.
3. The middle term must be twice the product of the terms a and b.

EXAMPLE 25   Factor $x^2 + 6x + 9$.

SOLUTION   The first term is a square, x^2. The last term is a square, 3^2. Finally, the middle term, $6x$, is twice the product of x and 3; thus, we can use the perfect-square property.

$$a^2 + 2ab + b^2 = (a + b)^2 \qquad x^2 + 6x + 9 = (x + 3)^2$$

EXAMPLE 26   The following examples also use this property.

(a) $x^2 - 14x + 49 = (x - 7)^2$

(b) $9k^2 + 12k + 4 = (3k + 2)^2$

(c) $100r^2 - 60r + 9 = (10r - 3)^2$

(d) $25u^2 + 40uv + 16v^2 = (5u + 4v)^2$

$\boxed{a^2}$ $\boxed{2ab}$ $\boxed{b^2}$ $\boxed{a}$ $\boxed{b}$

YES	NO
$x^2 + 8x + 16 = (x + 4)^2$	$x^2 + 4x + 16 = (x + 4)^2$

The middle term must be *twice* the product of the terms in the binomial.

PROBLEM SET 4.5

Factor the following polynomials completely.

See Example 21

1. $x^2 - 1$ **2.** $a^2 - 36$

3. $p^2 - 49$ **4.** $q^2 - 100$

5. $t^2 - 4$ **6.** $k^2 - 9$

7. $m^2 - 121$ **8.** $b^2 - 64$

See Examples 22 and 23

9. $4x^2 - 1$ **10.** $9y^2 - 16$

11. $16p^2 - 25$ **12.** $100r^2 - 49$

13. $81t^2 - 64$ **14.** $4 - 49x^2$

15. $9y^2 - 25$ **16.** $4 - 121m^2$

17. $49k^2 - 121$ **18.** $144u^2 - 25$

19. $4a^2 - 81$ **20.** $36m^2 - 49$

See Example 24

21. $x^4 - 1$ **22.** $a^4 - 81$

23. $16u^4 - 1$ **24.** $81k^4 - 256$

25. $256 - y^4$ **26.** $81 - 16r^4$

See Example 25

27. $x^2 + 2x + 1$ **28.** $a^2 - 4a + 4$

29. $b^2 - 18b + 81$ **30.** $w^2 + 12w + 36$

31. $y^2 + 14y + 49$ **32.** $z^2 - 10z + 25$

See
Example 26

33. $4x^2 + 20x + 25$

34. $4m^2 - 12m + 9$

35. $9a^2 - 42a + 49$

36. $25v^2 - 30v + 9$

37. $16u^2 + 24u + 9$

38. $25p^2 + 40p + 16$

39. $36m^2 - 12mn + n^2$

40. $36r^2 + 60rs + 25s^2$

Life science
application

41. A certain bacteria population grows according to the formula

$$P = 5000t^2 + 30,000t + 45,000$$

where t is the time in hours. Factor the polynomial on the right completely.

Physical science
application

42. In 2 seconds, an object drops 64 feet. In t seconds, it drops $16t^2$. A physicist studying the difference between the drop in t seconds and 2 seconds arrives at the relation

$$\text{difference} = 16t^2 - 64$$

Factor the expression on the right completely.

4.6   QUADRATIC EQUATIONS

As an application of factoring, we consider **quadratic equations,** which are second-degree equations with a **standard form**

$$ax^2 + bx + c = 0$$

where $a\,(\neq 0)$, b, and c are real numbers, and x is the variable to be solved for. To solve these, we use a very important property of zero.

PROPERTY 3 (**Zero-Product Law**)   Let M and N be real numbers. If $M \cdot N = 0$, then either $M = 0$, or $N = 0$, or both.

We use this property when we have an equation that is factored into a product equal to zero. Then either of these factors can be set equal to zero.

EXAMPLE 27   Solve $(x - 7)(2x + 1) = 0$.

Here we have a product equal to zero. By the zero-product law, either of these factors might be zero.

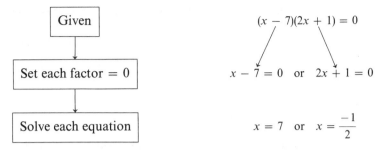

We get *two* solutions: $\left\{7, \dfrac{-1}{2}\right\}$. We can check that both of these are solutions by substituting into the original equation:

$$x = 7: \quad (7 - 7)(2 \cdot 7 + 1) = 0(15) = 0 \qquad \textit{Checks.}$$

$$x = \frac{-1}{2}: \quad \left(\frac{-1}{2} - 7\right)\left[2\left(\frac{-1}{2}\right) + 1\right] = -7\tfrac{1}{2}(0) = 0 \qquad \textit{Checks.}$$

In general, we have the following procedures for solving quadratic equations by factoring.

> To solve quadratic equations by factoring:
>
> **1.** Put the equation into standard form: $ax^2 + bx + c = 0$.
> **2.** Factor the polynomial completely.
> **3.** Set each factor equal to zero.
> **4.** Solve each of the resulting equations.
> **5.** Check the solutions.

EXAMPLE 28   Solve $x^2 - 3x - 10 = 0$.

SOLUTION   This is a quadratic equation in standard form. We factor this trinomial and set each factor to zero.

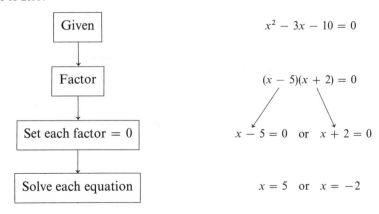

The solution set is $\{5, -2\}$. We can check these in the original equation:

$$x = 5: \quad 5^2 - 3(5) - 10 = 25 - 15 - 10 = 0 \qquad \textit{Checks.}$$
$$x = -2: \quad (-2)^2 - 3(-2) - 10 = 4 + 6 - 10 = 0 \qquad \textit{Checks.}$$

EXAMPLE 29   Solve $6t^2 = 15 - t$.

SOLUTION   We must first put this into standard form with the 0 on the right. Then, by trial and error, we factor the polynomial.

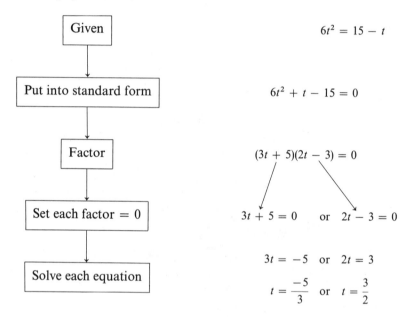

| Given | | $6t^2 = 15 - t$ |

| Put into standard form | | $6t^2 + t - 15 = 0$ |

| Factor | | $(3t + 5)(2t - 3) = 0$ |

| Set each factor $= 0$ | | $3t + 5 = 0 \quad$ or $\quad 2t - 3 = 0$ |

| Solve each equation | | $3t = -5 \quad$ or $\quad 2t = 3$ |
| | | $t = \dfrac{-5}{3} \quad$ or $\quad t = \dfrac{3}{2}$ |

Thus, the solution set is $\left\{\dfrac{-5}{3}, \dfrac{3}{2}\right\}$.

YES	NO
$(x + 1)(x + 2) = 0$	$(x + 1)(x + 2) = 5$
$(x + 1) = 0$ or $(x + 2) = 0$	$(x + 1) = 5$ or $(x + 2) = 5$

The zero-product law works only with a product equal to *zero*.

EXAMPLE 30   Solve $x^3 - 4x = 0$.

SOLUTION   This is a third-degree equation. But we can factor this and solve it.

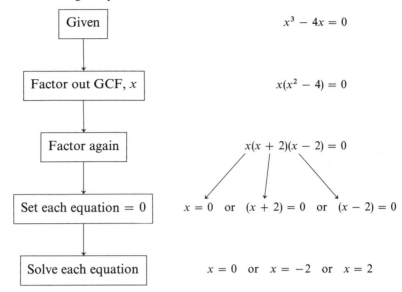

| Given | $x^3 - 4x = 0$ |

| Factor out GCF, x | $x(x^2 - 4) = 0$ |

| Factor again | $x(x + 2)(x - 2) = 0$ |

| Set each equation $= 0$ | $x = 0$ or $(x + 2) = 0$ or $(x - 2) = 0$ |

| Solve each equation | $x = 0$ or $x = -2$ or $x = 2$ |

There are three solutions, $\{0, -2, 2\}$.

PROBLEM SET 4.6

Solve the following equations.

See
Example 27

1. $(x + 1)(x - 2) = 0$ **2.** $(a - 3)(a - 4) = 0$

3. $(r + 5)(r + 6) = 0$ **4.** $(u - 7)(u - 8) = 0$

5. $(k + 7)(2k + 8) = 0$ **6.** $(y - 9)(3y + 1) = 0$

7. $(2z + 9)(3z + 2) = 0$ **8.** $(5b - 1)(4b + 3) = 0$

9. $m(8m + 5)(5m - 4) = 0$ **10.** $t(4t - 1)(3t + 13) = 0$

See
Examples
28 and 29

11. $x^2 + 4x + 3 = 0$ **12.** $a^2 + 7a + 10 = 0$

13. $p^2 + 2p - 8 = 0$ **14.** $k^2 - 9k + 14 = 0$

15. $m^2 - 4m - 21 = 0$ **16.** $t^2 - t - 30 = 0$

17. $b^2 - 4b - 45 = 0$ **18.** $y^2 - 14y + 48 = 0$

19. $2x^2 + 11x + 5 = 0$ **20.** $3a^2 + a - 10 = 0$

21. $2k^2 - 11k - 21 = 0$ **22.** $10m^2 + 39m + 14 = 0$

23. $15t^2 - 23t + 6 = 0$ **24.** $8n^2 + 2n - 21 = 0$

25. $5z^2 + 33z - 14 = 0$ **26.** $12r^2 - 23r + 10 = 0$

See
Example 29

27. $x^2 + 7x = -10$ **28.** $a^2 + 6a = 16$

29. $y^2 = y + 12$ **30.** $t^2 = 9t - 20$

31. $z^2 + 8z = -7$ **32.** $k^2 + k = 42$

33. $2x^2 = 5 + 9x$ **34.** $3m^2 = 22m + 16$

35. $4z^2 + 15 = 16z$ **36.** $3r^2 + 22r = -35$

37. $4q^2 = 7 - 27q$ **38.** $2u^2 + 40 = 21u$

See
Example 30

39. $x^3 - x = 0$ **40.** $a^3 - 9a = 0$

41. $y^3 = 4y$ **42.** $u^4 = u^2$

43. $9t^4 - 4t^2 = 0$ **44.** $16k^4 - 25k^2 = 0$

45. $m^3 + 9m^2 + 20m = 0$ **46.** $u^3 - 5u^2 - 24u = 0$

*Life science
application*   **47.** The yield Y in bushels per acre in a certain rice field is given by $Y = 29x - x^2$, where x is the density of the rice plant. Find what densities produce a yield of $Y = 100$ by solving

$$100 = 29x - x^2$$

*Business
application*   **48.** The *marginal profit MP* for producing x shirts (in thousands) is given by $MP = -x^2 + 40x - 300$. Find the number of shirts that give the greatest profit by solving $MP = -x^2 + 40x - 300 = 0$. (Consider only the smaller of the solutions.)

*Physical science
application*   **49.** The height of a thrown object is given by $h = -16t^2 + 64t$. Find the times at which the height is 48 by solving

$$-16t^2 + 64t = 48$$

4.7   WORD PROBLEMS

In this section we study the word problems that lead to quadratic equations. We follow the same word-problem approach that we used in Chapter 2.

> To solve word problems that lead to quadratic equations:
>
> 1. Identify the unknowns.
> 2. Translate the given information into an equation.
> 3. Put the equation into standard form.
> 4. Solve the equation.
> 5. Check to see that both answers make sense in the original problem.

EXAMPLE 31   One positive number is 3 more than another. Their product is 54. Find the numbers.

SOLUTION   We identify the unknowns as x and $x + 3$. Then we translate and solve.

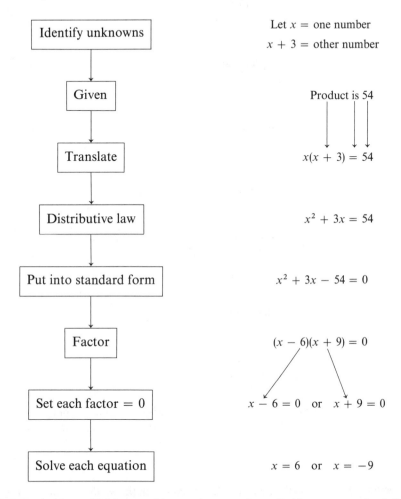

Let x = one number
$x + 3$ = other number

Product is 54

$x(x + 3) = 54$

$x^2 + 3x = 54$

$x^2 + 3x - 54 = 0$

$(x - 6)(x + 9) = 0$

$x - 6 = 0$ or $x + 9 = 0$

$x = 6$ or $x = -9$

Since the problem asked for a positive number, we reject -9. Thus, one number is 6, and the other is 3 more, or 9. We check that their product is 54.

EXAMPLE 32   The sum of two numbers is 10. The sum of their squares is 52. Find the numbers.

SOLUTION   We first identify the unknowns as x and $10 - x$.

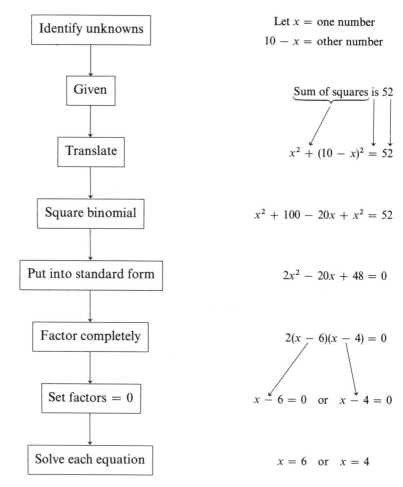

| Identify unknowns | Let x = one number |
| | $10 - x$ = other number |

Given — Sum of squares is 52

Translate — $x^2 + (10 - x)^2 = 52$

Square binomial — $x^2 + 100 - 20x + x^2 = 52$

Put into standard form — $2x^2 - 20x + 48 = 0$

Factor completely — $2(x - 6)(x - 4) = 0$

Set factors = 0 — $x - 6 = 0$ or $x - 4 = 0$

Solve each equation — $x = 6$ or $x = 4$

The numbers are 4 and 6. Their sum is 10, and the sum of their squares is $4^2 + 6^2 = 16 + 36 = 52$. Notice that when we had the equation $2(x - 6)(x - 4) = 0$ we did not bother to set $2 = 0$. We know that this is not true.

EXAMPLE 33   The length of a rectangle is 5 more than the width. The area of the rectangle is 36. Find the dimensions.

x | (rectangle) | $x+5$

SOLUTION   Recall that the area of a rectangle is length times width.

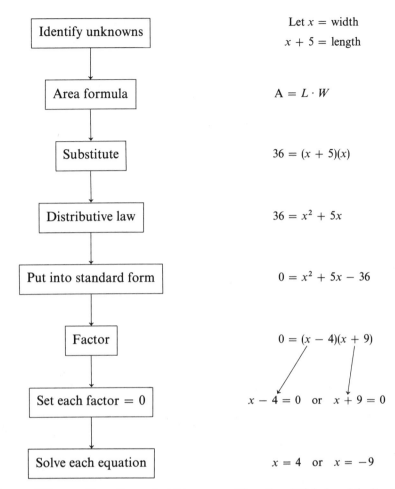

Since a width cannot be -9, we reject this answer. Thus, the width is 4, and the length is 9. (Notice that here it was convenient to put all the terms on the right. This is acceptable.)

PROBLEM SET 4.7

Solve the following word problems.

See
Example 31

1. The sum of two numbers is 12, and their product is 32. Find the numbers.

2. The sum of two numbers is 15, and their product is 36. Find the numbers.

3. One positive number is 7 more than another number. Their product is 18. Find the numbers.

4. One positive number is 8 more than another number. Their product is 48. Find the numbers.

5. The product of two consecutive positive integers is 72. Find the numbers.

6. The product of two consecutive even positive integers is 48. Find the numbers.

7. The product of two consecutive positive odd integers is 143. Find the numbers.

8. The product of two consecutive positive integers is 110. Find the numbers.

See Example 32 **9.** The sum of two integers is 12. The sum of their squares is 122. Find the numbers.

10. The sum of two integers is 15. The sum of their squares is 125. Find the numbers.

11. The sum of the squares of two consecutive positive integers is 145. Find the numbers.

12. The sum of the squares of two consecutive positive even integers is 100. Find the numbers.

13. The square of a positive number is 14 more than 5 times the number. Find the number.

14. The square of a positive number is 20 less than 12 times the number. Find the number.

See Example 33 **15.** The length of a rectangle is 6 more than the width. The area is 40. Find the dimensions.

16. The width of a rectangle is 10 less than the length. The area is 75. Find the dimensions.

17. The perimeter of a certain rectangle is 24. The area is 32. Find the dimensions.

18. The perimeter of a certain rectangle is 26. The area is 30. Find the dimensions.

CHAPTER 4 SUMMARY

Important Words and Phrases

composite (4.1)

difference of squares (4.5)

factoring (4.1)

greatest common factor (GCF) (4.2)

perfect-square trinomial (4.5)

prime (4.1)

quadratic equation (4.6)

standard form (4.6)

zero-product law (4.6)

Important Properties

Every composite can be written as a product of primes.

A polynomial factors into the product of:
(a) The GCF of all its terms, and
(b) The polynomial divided by this GCF.

For any real numbers a and b,

$$a^2 - b^2 = (a + b)(a - b)$$
$$a^2 + 2ab + b^2 = (a + b)^2$$
$$a^2 - 2ab + b^2 = (a - b)^2$$

If $a \cdot b = 0$, then either $a = 0$, $b = 0$, or both.

Important Procedures

To factor a trinomial of the type $x^2 + Mx + N$:
1. Find all pairs of factors of N.
2. Choose the pair of factors, a and b (if it exists), whose sum is M.
3. The factoring is $(x + a)(x + b)$.

To factor a trinomial of the type $ax^2 + bx + c$:
1. Factor the first term, ax^2, and the last term, c.
2. Look at all binomial combinations with these factors.
3. Find the combination (if it exists) that gives the middle term, bx.
4. If a combination gives $-bx$, reverse both signs of the factors of c.

To solve quadratic equations:
1. Put the equation into standard form, $ax^2 + bx + c = 0$.
2. Factor the polynomial completely.
3. Set each factor equal to zero.
4. Solve each of the resulting equations.
5. Check the solutions in the original equation.

CHAPTER 4 REVIEW EXERCISES

For the following numbers, identify the primes and factor the composites into primes.

1. 71   **2.** 81   **3.** 91

For the following sets, find the greatest common factor (GCF).

4. 8, 12, 18

5. $2m^2n^4$, $4m^4n^3$, $6mn^5$

Factor the following polynomials completely.

6. $10x - 15y$

7. $4x^2 - 6x^5$

8. $6a^2b^4 - 12a^4b^3 + 24ab^5$

9. $25u^2v^3w^4t^5 - 35u^3v^3w^3t^3$

10. $x^2 + 11x + 18$

11. $y^2 - 5y - 24$

12. $k^2 - 11k + 10$

13. $2a^3 - 6a^2 - 56a$

14. $2x^2 + 13x + 15$

15. $12r^2 - 5r - 3$

16. $6u^2 - 19u + 10$

17. $4p^4 + 17p^3 - 15p^2$

18. $x^2 - 64$

19. $4 - 81p^2$

20. $16q^4 - 81$

21. $k^2 + 4k + 4$

22. $9m^2 - 30m + 25$

Solve the following equations.

23. $(a + 7)(a - 4) = 0$

24. $p^2 - 7p - 8 = 0$

25. $6x^2 + 7x - 5 = 0$

26. $x^2 + 2x = 35$

27. $x^3 - 25x = 0$

Solve the following word problems.

28. The product of two consecutive positive even integers is 120. Find the integers.

29. The sum of two numbers is 11. The sum of their squares is 85. Find the numbers.

30. The length of a rectangle is 6 more than its width. The area is 55. Find the dimensions.

5
FRACTIONS

5.1 BASIC IDEAS

In grade school we learned that a fraction was a portion of a whole. Usually, we saw a cut-up pie displaying a fraction. More generally, a **fraction,** or **rational number,** is the quotient $\dfrac{a}{b}$ of any two integers a and b (where $b \neq 0$). We call the top number the **numerator** and the bottom number the **denominator.** (Sometimes, we simply call these the *top* and *bottom*.) As examples,

$$\frac{1}{2}, \quad \frac{-3}{4}, \quad \frac{8}{-4}, \quad \frac{15}{93}, \quad \frac{0}{7}, \quad \frac{10}{10}$$

are all fractions, or rational numbers.

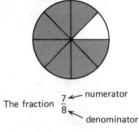

The fraction $\dfrac{7 \leftarrow \text{numerator}}{8 \leftarrow \text{denominator}}$

A special class of fractions is the type $\dfrac{m}{m}$ (where $m \neq 0$). Since the fraction $\dfrac{m}{m} = m \div m = 1$, we have the following property.

PROPERTY 1 For any real number m ($\neq 0$),

$$\frac{m}{m} = 1$$

For example, $\dfrac{10}{10} = \dfrac{3}{3} = \dfrac{23}{23} = \dfrac{-7}{-7} = \dfrac{-219}{-219} = 1$. Recall the following rule for multiplying fractions.

> **PROPERTY 2**   For any real numbers a, b, c, and d,
>
> $$\frac{a}{b} \cdot \frac{c}{d} = \frac{ac}{bd} \quad \text{(where } b \neq 0 \text{ and } d \neq 0)$$

We use these two facts to rewrite fractions in simpler or in built-up forms. We also use the fact that $k \cdot 1 = k$, for any real number k. Consider $\dfrac{15}{21}$.

$$\frac{15}{21} = \frac{5 \cdot 3}{7 \cdot 3} = \frac{5}{7} \cdot \frac{3}{3} = \frac{5}{7} \cdot 1 = \frac{5}{7}$$

A fraction, such as $\dfrac{5}{7}$ above, is said to be **reduced to simplest terms** if the numerator and denominator have no common factors. We can use the following property to help us reduce fractions.

> **PROPERTY 3**   For any real numbers A, B, C, and K (B, C, K not zero),
>
> $$\frac{A}{B} = \frac{A}{B} \cdot \frac{C}{C} = \frac{AC}{BC} \quad \text{and} \quad \frac{A}{B} = \frac{A \div K}{B \div K}$$

In words, *we can multiply or divide the top and bottom of a fraction by the same number, and the fraction keeps the same value.*

EXAMPLE 1   Reduce $\dfrac{16}{60}$ to simplest terms.

SOLUTION   We factor the top and bottom and look for common factors.

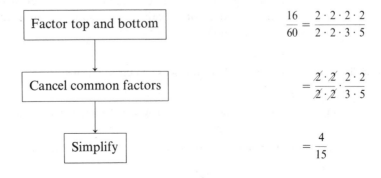

EXAMPLE 2   Reduce $\dfrac{66}{88}$ to simplest terms.

SOLUTION   We factor the top and bottom completely.

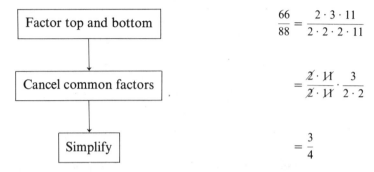

Let us summarize our general procedure for reducing.

> To reduce a fraction to simplest terms:
>
> **1.** Factor the top and bottom completely.
>
> **2.** Cancel all common factors.
>
> **3.** Simplify the remaining terms.

An **algebraic fraction,** or **rational expression,** is the quotient of two poly-nomials, $\dfrac{P}{Q}$, where $Q \neq 0$. As examples,

$$\frac{2x}{y+3}, \quad \frac{4ab}{5cd}, \quad \frac{a+b}{c+d}, \quad \frac{x^2+5x+6}{2x^2-4x+7}$$

are algebraic fractions. We have to be careful with the values that we substitute into the fraction (since we cannot divide by 0). The set of values that we can substitute into the fraction is called its **domain.** This is generally the set of points where the denominator is not zero.

EXAMPLE 3   The following are algebraic fractions and their domains.

(a) $\dfrac{3}{x}$; domain = {all reals, $x \neq 0$}. (If $x = 0$, the denominator is 0.)

(b) $\dfrac{a+2}{a-5}$; domain = {all reals, $a \neq 5$}. ($a = 5$ makes the denominator 0.)

(c) $\dfrac{t}{5}$; domain = {all reals}. (The denominator is never 0.)

(d) $\dfrac{m}{m^2 - 9}$; domain = {all reals, $m \neq -3, 3$}. (3 and -3 make the bottom 0.)

    We reduce algebraic fractions the same way that we do with numerical fractions: *factor the top and bottom completely, cancel, and simplify.*

EXAMPLE 4   Reduce $\dfrac{4a^2x^5}{10a^3x^3}$.

SOLUTION   This is very similar to the quotients of monomials that we saw in Chapter 3. We reduce this by factoring.

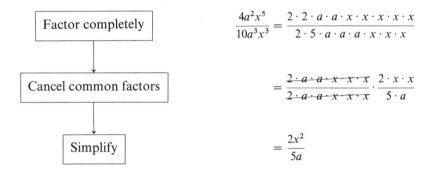

Notice that what we are really doing is finding a fraction equal to 1, and "canceling" it.

EXAMPLE 5   Reduce $\dfrac{x^2 + 5x}{3x + 15}$.

SOLUTION   We factor the top and bottom completely.

$$\boxed{\text{Factor top and bottom}} \qquad \frac{x^2 + 5x}{3x + 15} = \frac{x(x + 5)}{3(x + 5)}$$

$$\boxed{\text{Cancel common terms}} \qquad = \frac{x}{3} \cdot \frac{\cancel{(x + 5)}}{\cancel{(x + 5)}} = \frac{x}{3}$$

EXAMPLE 6   Reduce $\dfrac{a^2 - b^2}{a^2 - ab}$.

SOLUTION   We factor the numerator and denominator completely.

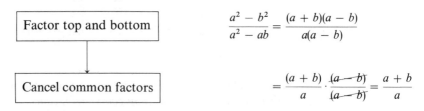

EXAMPLE 7   Reduce $\dfrac{6a - 15}{10a - 4a^2}$.

SOLUTION   After we factor, we have to rewrite this slightly in order to cancel.

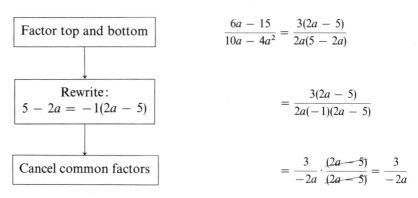

Notice that we rewrote $(5 - 2a)$ as $(-1)(2a - 5)$ so that it would cancel with the $(2a - 5)$ in the numerator.

EXAMPLE 8   Reduce $\dfrac{x^2 + x - 12}{x^2 - 8x + 15}$.

SOLUTION   We first factor completely.

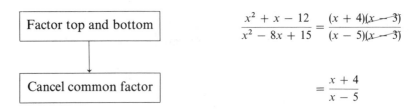

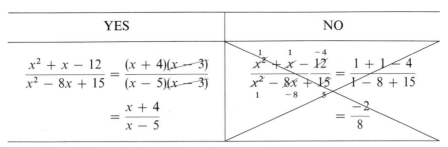

We can only cancel common *factors*.

PROBLEM SET 5.1

Reduce the following fractions to simplest terms.

See
Examples
1 and 2

1. $\dfrac{15}{50}$ **2.** $\dfrac{12}{32}$

3. $\dfrac{10}{28}$ **4.** $\dfrac{18}{21}$

5. $\dfrac{24}{30}$ **6.** $\dfrac{42}{36}$

7. $\dfrac{35}{65}$ **8.** $\dfrac{40}{85}$

9. $\dfrac{60}{64}$ **10.** $\dfrac{80}{96}$

11. $\dfrac{108}{72}$ **12.** $\dfrac{48}{64}$

13. $\dfrac{144}{132}$ **14.** $\dfrac{98}{96}$

Give the domain for each of the following algebraic fractions.

See
Example 3

15. $\dfrac{6}{a}$ **16.** $\dfrac{-7}{t}$

17. $\dfrac{4}{x-7}$ **18.** $\dfrac{18}{a-8}$

19. $\dfrac{x+9}{x+5}$ **20.** $\dfrac{x-12}{x+12}$

21. $\dfrac{m + 2}{5}$

22. $\dfrac{z - 7}{12}$

23. $\dfrac{1}{x^2 - 4}$

24. $\dfrac{5}{r^2 - 25}$

Reduce each of the following algebraic fractions to simplest terms.

See Example 4 **25.** $\dfrac{25a^2}{15a^4}$

26. $\dfrac{14x^2y}{21xy^5}$

27. $\dfrac{16rs^2}{24r^2s}$

28. $\dfrac{10m^2n^4}{15m^5n^3}$

29. $\dfrac{30a^2b^3}{25a^3b^2}$

30. $\dfrac{-36x^2y^2z^2}{40xy^2z^3}$

31. $\dfrac{32p^2q^4r^6}{28p^5q^3r}$

32. $\dfrac{12mk^2n^5}{20m^4kn^7}$

See Example 5 **33.** $\dfrac{7x + 14}{8x + 16}$

34. $\dfrac{5a - 15}{7a - 21}$

35. $\dfrac{b^2 + 4b}{2b + 8}$

36. $\dfrac{3t - 21}{t^2 - 7t}$

37. $\dfrac{3k - 9}{3k + 15}$

38. $\dfrac{2x - 12}{4x - 2}$

39. $\dfrac{x^2 + x}{x^2 - 3x}$

40. $\dfrac{2m^3 - 5m^2}{m^3 + 3m^2}$

See Example 6 **41.** $\dfrac{a^2 - 9}{a^2 - 3a}$

42. $\dfrac{t^2 - 25}{5t - 25}$

43. $\dfrac{x + 5}{x^2 - 25}$

44. $\dfrac{r + 6}{r^2 - 36}$

45. $\dfrac{x^2 + 4x - 5}{8x + 40}$

46. $\dfrac{x^2 - 9x + 14}{5x - 35}$

47. $\dfrac{xy + y^2}{xy - y^2}$

48. $\dfrac{ab - ax}{ab + ax}$

See Example 7 **49.** $\dfrac{2x - 8}{12 - 3x}$

50. $\dfrac{2a - 10}{20 - 4a}$

51. $\dfrac{a - b}{b - a}$

52. $\dfrac{6 - x}{x - 6}$

53. $\dfrac{16 - r^2}{r - 4}$

54. $\dfrac{t - 8}{64 - t^2}$

55. $\dfrac{9 - x^2}{x^2 - 3x}$

56. $\dfrac{2a^2 - 10a}{25 - a^2}$

See Example 8

57. $\dfrac{x^2 - 10x + 24}{x^2 - 9x + 20}$

58. $\dfrac{a^2 + 15a + 56}{a^2 + 14a + 48}$

59. $\dfrac{m^2 + m - 12}{m^2 - 8m + 15}$

60. $\dfrac{k^2 + 4k - 12}{k^2 + 11k + 30}$

61. $\dfrac{t^2 + 2t - 35}{t^2 - 25}$

62. $\dfrac{r^2 + 5r - 6}{r^2 - 36}$

63. $\dfrac{3x^2 - 17x + 10}{3x^2 + 19x - 14}$

64. $\dfrac{2k^2 + 11k + 5}{3k^2 + 16k + 5}$

Physical science application

65. In 3 seconds, an object drops 144 feet. In t seconds, it drops $16t^2$ feet. The average speed between 3 and t seconds is given by

$$\text{average speed} = \frac{16t^2 - 144}{t - 3}$$

Reduce this fraction completely.

Business application

66. The cost C to produce x units of a certain lamp is given by

$$C = -x^3 - 100x^2 + 2000x$$

The average cost is given by

$$\text{average cost} = \frac{C}{x} = \frac{-x^3 - 100x^2 + 2000x}{x}$$

Reduce this fraction completely.

5.2   MULTIPLICATION OF FRACTIONS

Multiplying number fractions and algebraic fractions is very similar. In fact, we use the same rule.

PROPERTY 4   For integers or polynomials A, B, C, and D (B and D not zero),

$$\frac{A}{B} \cdot \frac{C}{D} = \frac{A \cdot C}{B \cdot D}$$

EXAMPLE 9   (a) $\dfrac{2}{5} \cdot \dfrac{3}{7} = \dfrac{2 \cdot 3}{5 \cdot 7} = \dfrac{6}{35}$

(b) $\dfrac{-4}{5} \cdot \dfrac{7}{8} = \dfrac{-4 \cdot 7}{5 \cdot 8} = \dfrac{-28}{40} = \dfrac{-7}{10}$

EXAMPLE 10   Multiply $\dfrac{-8}{9} \cdot \dfrac{15}{16}$.

SOLUTION   When we multiply number fractions, it is usually easier to cancel before we multiply.

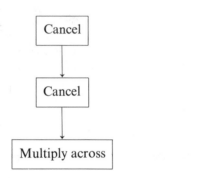

With algebraic fractions, we have to factor first in order to cancel.

> To multiply algebraic fractions:
>
> **1.** Factor all numerators and denominators completely.
>
> **2.** Multiply across: numerator times numerator, denominator times denominator.
>
> **3.** Cancel common factors and simplify.

EXAMPLE 11   Multiply $\dfrac{2a^2}{3b^3} \cdot \dfrac{9b^4}{8a^5}$.

SOLUTION   We write this out in "slow motion."

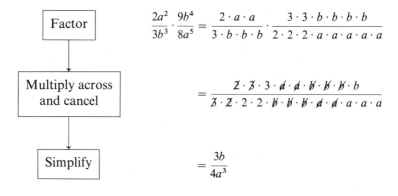

$$\boxed{\text{Factor}} \qquad \frac{2a^2}{3b^3} \cdot \frac{9b^4}{8a^5} = \frac{2 \cdot a \cdot a}{3 \cdot b \cdot b \cdot b} \cdot \frac{3 \cdot 3 \cdot b \cdot b \cdot b \cdot b}{2 \cdot 2 \cdot 2 \cdot a \cdot a \cdot a \cdot a \cdot a}$$

$$\boxed{\substack{\text{Multiply across} \\ \text{and cancel}}} \qquad = \frac{\cancel{2} \cdot \cancel{3} \cdot 3 \cdot \cancel{a} \cdot \cancel{a} \cdot \cancel{b} \cdot \cancel{b} \cdot \cancel{b} \cdot b}{\cancel{3} \cdot \cancel{2} \cdot 2 \cdot 2 \cdot \cancel{b} \cdot \cancel{b} \cdot \cancel{b} \cdot \cancel{a} \cdot \cancel{a} \cdot a \cdot a \cdot a}$$

$$\boxed{\text{Simplify}} \qquad = \frac{3b}{4a^3}$$

EXAMPLE 12   Multiply $\dfrac{x^2 + 5x}{x^2 + 6x + 5} \cdot \dfrac{x^4 + x^3}{x^3 + 6x^2}$.

SOLUTION   We begin by factoring every expression.

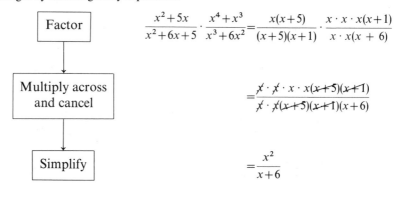

$$\boxed{\text{Factor}} \qquad \frac{x^2+5x}{x^2+6x+5} \cdot \frac{x^4+x^3}{x^3+6x^2} = \frac{x(x+5)}{(x+5)(x+1)} \cdot \frac{x \cdot x \cdot x(x+1)}{x \cdot x(x+6)}$$

$$\boxed{\substack{\text{Multiply across} \\ \text{and cancel}}} \qquad = \frac{\cancel{x} \cdot \cancel{x} \cdot x \cdot x\cancel{(x+5)}\cancel{(x+1)}}{\cancel{x} \cdot \cancel{x}\cancel{(x+5)}\cancel{(x+1)}(x+6)}$$

$$\boxed{\text{Simplify}} \qquad = \frac{x^2}{x+6}$$

EXAMPLE 13   Multiply $\dfrac{a^2 - a - 12}{a^2 + 11a + 30} \cdot \dfrac{a^2 + 13a + 40}{a^2 + 4a - 32}$.

SOLUTION   We begin by factoring all the polynomials involved.

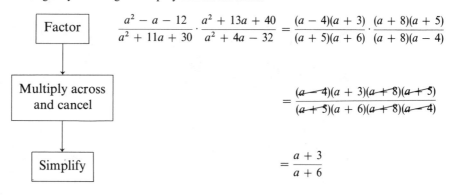

$$\boxed{\text{Factor}} \qquad \frac{a^2 - a - 12}{a^2 + 11a + 30} \cdot \frac{a^2 + 13a + 40}{a^2 + 4a - 32} = \frac{(a-4)(a+3)}{(a+5)(a+6)} \cdot \frac{(a+8)(a+5)}{(a+8)(a-4)}$$

$$\boxed{\substack{\text{Multiply across} \\ \text{and cancel}}} \qquad = \frac{\cancel{(a-4)}(a+3)\cancel{(a+8)}\cancel{(a+5)}}{\cancel{(a+5)}(a+6)\cancel{(a+8)}\cancel{(a-4)}}$$

$$\boxed{\text{Simplify}} \qquad = \frac{a+3}{a+6}$$

PROBLEM SET 5.2

Multiply the following fractions and reduce to simplest terms.

See
Examples
9 and 10

1. $\dfrac{2}{3} \cdot \dfrac{4}{11}$

2. $\dfrac{-3}{7} \cdot \dfrac{4}{13}$

3. $\dfrac{1}{2} \cdot \dfrac{1}{4}$

4. $\dfrac{1}{3} \cdot \dfrac{-1}{9}$

5. $\dfrac{2}{3} \cdot \dfrac{6}{11}$

6. $\dfrac{-4}{5} \cdot \dfrac{15}{22}$

7. $\dfrac{-5}{8} \cdot \dfrac{2}{15}$

8. $\dfrac{-10}{21} \cdot \dfrac{-7}{15}$

9. $\dfrac{14}{15} \cdot \dfrac{20}{21}$

10. $\dfrac{36}{25} \cdot \dfrac{35}{18}$

11. $\dfrac{30}{49} \cdot \dfrac{28}{27}$

12. $\dfrac{60}{77} \cdot \dfrac{14}{25}$

See
Example 11

13. $\dfrac{3x^2}{5y^3} \cdot \dfrac{10y^4}{9x^6}$

14. $\dfrac{4a^4}{5b^3} \cdot \dfrac{15b^7}{2a^5}$

15. $\dfrac{5mn^2}{7p^2q^2} \cdot \dfrac{14p^3q}{15m^2n^4}$

16. $\dfrac{6u^2v^4}{11w^3t} \cdot \dfrac{22wt^5}{21u^5v}$

17. $\dfrac{ab^2c^3}{6x^2y^3z^5} \cdot \dfrac{3x^5yz^2}{4a^4b^4c^4}$

18. $\dfrac{m^2n^4k^7}{15u^2v^7w^4} \cdot \dfrac{5u^8v^6w^2}{6m^5n^5k^5}$

See
Example 12

19. $\dfrac{a^2 - 25}{a^2} \cdot \dfrac{a^4}{a^2 + 5a}$

20. $\dfrac{x}{x^2 - 49} \cdot \dfrac{x^3 + 7x^2}{x^5}$

21. $\dfrac{k + 6}{k + 2} \cdot \dfrac{k^2 - 4}{k^2 - 36}$

22. $\dfrac{a - 3}{a + 1} \cdot \dfrac{a^2 - 1}{a^2 - 9}$

23. $\dfrac{r^2 + 6r}{r^2 - 3r} \cdot \dfrac{2r - 6}{3r + 18}$

24. $\dfrac{3k - 6}{2k^2 + k} \cdot \dfrac{6k + 3}{k^3 - 2k^2}$

See
Example 13

25. $\dfrac{x^7 - 6x^6}{x^2 - x - 30} \cdot \dfrac{x^2 - 25}{x^4 - 36x^2}$

26. $\dfrac{r^2 + 6r - 16}{r^2 + 8r - 20} \cdot \dfrac{r^2 + 6r - 40}{r^2 + 3r - 40}$

27. $\dfrac{m^2 - 16m + 63}{m^2 - 49} \cdot \dfrac{m^2 + 12m + 35}{m^2 - 25}$

28. $\dfrac{t^2 - t - 20}{t^2 - 16} \cdot \dfrac{t^2 + 2t - 24}{t^2 - 36}$

29. $\dfrac{3u^2 - 14u - 5}{u^2 - 3u - 10} \cdot \dfrac{u^2 + 11x + 10}{3u^2 + 31u + 10}$

30. $\dfrac{4x^2 - 21x - 18}{4x^2 + 23x + 15} \cdot \dfrac{x^2 - 25}{x^2 - 7x + 6}$

5.3   DIVISION OF FRACTIONS

Division of fractions is very much related to the multiplication of fractions. The key to division of fractions is the **reciprocal** (or **multiplicative inverse**). Recall from Chapter 1 that every real number x (except 0) has a reciprocal $\dfrac{1}{x}$ that satisfies the following relation.

PROPERTY 5   For any real number x ($x \neq 0$),

$$x \cdot \frac{1}{x} = 1$$

EXAMPLE 14   The following are examples of fractions, their reciprocals, and their products. (Notice that the product is always 1.)

	NUMBER	RECIPROCAL	PRODUCT
(a)	10	$\dfrac{1}{10}$	$\dfrac{10}{1} \cdot \dfrac{1}{10} = 1$
(b)	$\dfrac{-2}{15}$	$\dfrac{-15}{2}$	$\dfrac{-2}{15} \cdot \dfrac{-15}{2} = 1$
(c)	$\dfrac{x}{y}$	$\dfrac{y}{x}$	$\dfrac{x}{y} \cdot \dfrac{y}{x} = 1$
(d)	$\dfrac{a+b}{c+d}$	$\dfrac{c+d}{a+b}$	$\dfrac{a+b}{c+d} \cdot \dfrac{c+d}{a+b} = 1$

To divide fractions:

1. Replace the divisor (second fraction) by its reciprocal. (Invert the divisor.)

2. Multiply the results.

This is a familiar rule: *Invert and multiply*. In symbols, this is given by the following property.

PROPERTY 6   For numbers or polynomials A, B, C, and D (B, C, and D not zero),

$$\frac{A}{B} \div \frac{C}{D} = \frac{A}{B} \cdot \frac{D}{C} = \frac{A \cdot D}{B \cdot C}$$

EXAMPLE 15   The following divisions use the *invert-and-multiply* rule.

(a) $\dfrac{2}{3} \div \dfrac{5}{7} = \dfrac{2}{3} \cdot \dfrac{7}{5} = \dfrac{14}{15}$

(b) $6 \div \dfrac{3}{4} = \dfrac{6}{1} \cdot \dfrac{4}{3} = \dfrac{24}{3} = 8$

(c) $\dfrac{2}{5} \div 10 = \dfrac{2}{5} \cdot \dfrac{1}{10} = \dfrac{2}{50} = \dfrac{1}{25}$

(d) $\dfrac{4}{9} \div \dfrac{8}{15} = \dfrac{\cancel{4}^{1}}{\cancel{9}_{3}} \cdot \dfrac{\cancel{15}^{5}}{\cancel{8}_{2}} = \dfrac{1}{3} \cdot \dfrac{5}{2} = \dfrac{5}{6}$

When dividing algebraic fractions, we use the same *invert and multiply* rule. After inverting, we factor the polynomials involved, cancel, and simplify.

EXAMPLE 16   Divide $\dfrac{4a^2}{5b^5} \div \dfrac{2a^3}{15b^2}$.

SOLUTION   We first invert the divisor. Then we factor and multiply.

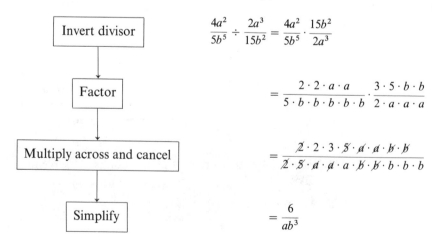

$$\frac{4a^2}{5b^5} \div \frac{2a^3}{15b^2} = \frac{4a^2}{5b^5} \cdot \frac{15b^2}{2a^3}$$

$$= \frac{2 \cdot 2 \cdot a \cdot a}{5 \cdot b \cdot b \cdot b \cdot b \cdot b} \cdot \frac{3 \cdot 5 \cdot b \cdot b}{2 \cdot a \cdot a \cdot a}$$

$$= \frac{\cancel{2} \cdot 2 \cdot 3 \cdot \cancel{5} \cdot \cancel{a} \cdot \cancel{a} \cdot \cancel{b} \cdot \cancel{b}}{\cancel{2} \cdot \cancel{5} \cdot \cancel{a} \cdot \cancel{a} \cdot a \cdot \cancel{b} \cdot \cancel{b} \cdot b \cdot b \cdot b}$$

$$= \frac{6}{ab^3}$$

When dividing monomials such as these, it may not be necessary to write out all the factors. This can often be done mentally. We do it here to emphasize the procedure.

EXAMPLE 17 Divide $\dfrac{a^2 + 5a}{a - 4} \div \dfrac{a^2 - 25}{a^2 - 16}$.

SOLUTION We first invert the divisor. Then we factor, multiply, and cancel.

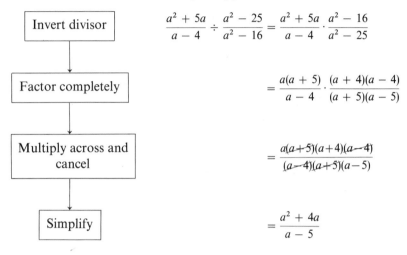

$$\boxed{\text{Invert divisor}} \qquad \frac{a^2 + 5a}{a - 4} \div \frac{a^2 - 25}{a^2 - 16} = \frac{a^2 + 5a}{a - 4} \cdot \frac{a^2 - 16}{a^2 - 25}$$

$$\boxed{\text{Factor completely}} \qquad = \frac{a(a + 5)}{a - 4} \cdot \frac{(a + 4)(a - 4)}{(a + 5)(a - 5)}$$

$$\boxed{\begin{array}{c}\text{Multiply across and}\\\text{cancel}\end{array}} \qquad = \frac{a\cancel{(a+5)}(a+4)\cancel{(a-4)}}{\cancel{(a-4)}\cancel{(a+5)}(a-5)}$$

$$\boxed{\text{Simplify}} \qquad = \frac{a^2 + 4a}{a - 5}$$

EXAMPLE 18 Divide $\dfrac{k^2 - 4k - 12}{k^2 - 3k - 18} \div \dfrac{k^2 - 6k - 16}{k^2 - 4k - 21}$.

SOLUTION Again, we invert, factor, multiply, cancel, and simplify.

$$\boxed{\begin{array}{c}\text{Invert}\\\text{divisor}\end{array}} \qquad \frac{k^2 - 4k - 12}{k^2 - 3k - 18} \div \frac{k^2 - 6k - 16}{k^2 - 4k - 21} = \frac{k^2 - 4k - 12}{k^2 - 3k - 18} \cdot \frac{k^2 - 4k - 21}{k^2 - 6k - 16}$$

$$\boxed{\begin{array}{c}\text{Factor}\\\text{completely}\end{array}} \qquad = \frac{(k - 6)(k + 2)}{(k - 6)(k + 3)} \cdot \frac{(k - 7)(k + 3)}{(k - 8)(k + 2)}$$

$$\boxed{\begin{array}{c}\text{Multiply}\\\text{and cancel}\end{array}} \qquad = \frac{\cancel{(k-6)}\cancel{(k+2)}(k - 7)\cancel{(k+3)}}{\cancel{(k-6)}\cancel{(k+3)}(k - 8)\cancel{(k+2)}}$$

$$\boxed{\text{Simplify}} \qquad = \frac{k - 7}{k - 8}$$

PROBLEM SET 5.3

Give the reciprocal for each of the following fractions.

See Example 14

1. $\dfrac{3}{7}$ **2.** $\dfrac{4}{9}$ **3.** $\dfrac{-5}{3}$

4. $\dfrac{6}{11}$ **5.** 12 **6.** -8

7. $\dfrac{u}{v}$ **8.** $\dfrac{5a}{-4b}$ **9.** $\dfrac{x^2 + y^2}{z}$

10. $\dfrac{a^3}{b + c + d}$ **11.** $\dfrac{k^2 - 25}{k^2 - 9}$ **12.** $\dfrac{a^2 + 2a + 1}{a^2 - 6a + 7}$

Divide the following fractions and reduce to simplest terms.

See Example 15

13. $\dfrac{3}{5} \div \dfrac{2}{7}$ **14.** $\dfrac{2}{11} \div \dfrac{4}{9}$

15. $\dfrac{2}{9} \div \dfrac{4}{3}$ **16.** $\dfrac{5}{7} \div \dfrac{10}{21}$

17. $\dfrac{-4}{9} \div \dfrac{8}{15}$ **18.** $\dfrac{10}{11} \div \dfrac{40}{33}$

19. $\dfrac{-24}{25} \div \dfrac{8}{15}$ **20.** $\dfrac{-32}{35} \div \dfrac{-16}{25}$

21. $\dfrac{21}{100} \div \dfrac{49}{50}$ **22.** $\dfrac{15}{14} \div \dfrac{25}{42}$

See Example 16

23. $\dfrac{a}{b^2} \div \dfrac{a^3}{b}$ **24.** $\dfrac{u^4}{v^3} \div \dfrac{u^3}{v^7}$

25. $\dfrac{4x^7 y}{9r^2 s^3} \div \dfrac{8x^3 y^3}{3r^4 s^2}$ **26.** $\dfrac{10a^2 b^3}{3m^2 n^5} \div \dfrac{5a^4 b^2}{9m^4 n^3}$

27. $\dfrac{a^2 b^3 c^4}{4x^2 y^3} \div \dfrac{3a^3 b^4}{8x^4 y^3 z}$ **28.** $\dfrac{u^7 v^3 w^4}{10r^2 s^9} \div \dfrac{2u^2 v^5}{5r^3 s^7 t^3}$

See Example 17

29. $\dfrac{x^2}{x^4 - 6x^3} \div \dfrac{x^4}{2x - 12}$ **30.** $\dfrac{r^4}{r^2 + 4r} \div \dfrac{r^7}{r^2 - 16}$

31. $\dfrac{a + 5}{2a - 6} \div \dfrac{a^2 + 5a}{5a - 15}$ **32.** $\dfrac{5x - 10}{4x + 12} \div \dfrac{x^2 - 4}{x^2 - 9}$

33. $\dfrac{m^2 + 5m}{m^2 - 25} \div \dfrac{2m + 6}{2m - 10}$

34. $\dfrac{k^3 - k^2}{k^2 - 1} \div \dfrac{3k + 9}{3k + 3}$

See
Example 18

35. $\dfrac{p^2 + 9p + 20}{p^2 + 4p - 21} \div \dfrac{p^2 + 7p + 10}{p^2 + 9p + 14}$

36. $\dfrac{a^2 - 2a - 35}{a^2 - 49} \div \dfrac{a^2 + 7a + 12}{a^2 + 2a - 8}$

37. $\dfrac{t^2 - 3t - 28}{t^2 - t - 42} \div \dfrac{t^2 - 100}{t^2 + 19t + 90}$

38. $\dfrac{x^2 + 8x + 15}{x^2 + 9x + 20} \div \dfrac{x^2 - 5x - 14}{x^2 + 6x + 8}$

39. $\dfrac{r^2 - 3r + 2}{5r^2 - 8r + 3} \div \dfrac{4r^2 + 25r - 21}{5r^2 + 32r - 21}$

40. $\dfrac{m^2 + m - 56}{3m^2 + 25m + 8} \div \dfrac{m^2 + 3m - 10}{3m^2 + 16m + 5}$

5.4 ADDITION AND SUBTRACTION OF FRACTIONS (LIKE DENOMINATORS)

When adding or subtracting fractions, we must have the *same denominator*. Consider $\dfrac{3}{11} + \dfrac{5}{11}$.

$$\frac{3}{11} + \frac{5}{11} = 3 \cdot \frac{1}{11} + 5 \cdot \frac{1}{11} = (3 + 5)\frac{1}{11}$$

$$= 8 \cdot \frac{1}{11} = \frac{8}{11}$$

Here we have used the distributive law to show that we add the numerators if the denominators are the same.

PROPERTY 7 For real numbers or polynomials A, B, and C $(C \neq 0)$,

$$\frac{A}{C} + \frac{B}{C} = \frac{A + B}{C} \quad \text{and} \quad \frac{A}{C} - \frac{B}{C} = \frac{A - B}{C}$$

We call C a **common denominator**.

EXAMPLE 19 The following additions each have a common denominator. To add the fractions, we simply add the numerators.

(a) $\dfrac{1}{7} + \dfrac{2}{7} = \dfrac{1+2}{7} = \dfrac{3}{7}$

(b) $\dfrac{1}{12} + \dfrac{-5}{12} = \dfrac{1+(-5)}{12} = \dfrac{-4}{12} = \dfrac{-1}{3}$

(c) $\dfrac{4}{13} + \dfrac{-1}{13} + \dfrac{2}{13} = \dfrac{4+(-1)+2}{13} = \dfrac{5}{13}$

To subtract fractions with a common denominator, we *subtract the numerators*. Often, this will mean changing the sign and adding.

EXAMPLE 20 The following subtractions each have a common denominator. We change each subtraction to addition of the opposite.

(a) $\dfrac{7}{11} - \dfrac{13}{11} = \dfrac{7}{11} \oplus \dfrac{\ominus 13}{11} = \dfrac{7+(-13)}{11} = \dfrac{-6}{11}$

(b) $\dfrac{1}{10} - \dfrac{-3}{10} = \dfrac{1}{10} \oplus \dfrac{\oplus 3}{10} = \dfrac{1+3}{10} = \dfrac{4}{10} = \dfrac{2}{5}$

(c) $\dfrac{4}{17} - \dfrac{7}{17} + \dfrac{11}{17} = \dfrac{4}{17} \oplus \dfrac{\ominus 7}{17} + \dfrac{11}{17} = \dfrac{4+(-7)+11}{17} = \dfrac{8}{17}$

For algebraic fractions with a *common denominator*, we follow the same procedures: add the numerators, or change subtraction to addition of the opposite.

EXAMPLE 21 The following additions each have a common denominator, so we add the numerators algebraically and simplify.

(a) $\dfrac{x}{a} + \dfrac{3}{a} = \dfrac{x+3}{a}$

(b) $\dfrac{x+5}{a+b} + \dfrac{2x-3}{a+b} = \dfrac{x+5+2x-3}{a+b} = \dfrac{3x+2}{a+b}$

(c) $\dfrac{2a-5}{(x-5)(x-3)} + \dfrac{4a+1}{(x-5)(x-3)} = \dfrac{6a-4}{(x-5)(x-3)}$

EXAMPLE 22 Subtract $\dfrac{4x-7}{a^2-b^2} - \dfrac{9x-3}{a^2-b^2}$.

SOLUTION We replace the subtraction by addition of the opposite by changing *all* the signs in the second numerator.

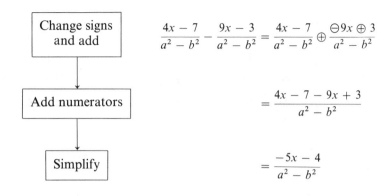

YES	NO
$-\dfrac{9x-3}{a^2-b^2} = \oplus\dfrac{\ominus9x\oplus3}{a^2-b^2}$	$-\dfrac{9x-3}{a^2-b^2} \oplus \dfrac{\ominus9x-3}{a^2-b^2}$

When subtracting, we change *all* the signs in the numerator.

PROBLEM SET 5.4

Simplify the following.

See Example 19

1. $\dfrac{1}{9} + \dfrac{4}{9}$

2. $\dfrac{1}{8} + \dfrac{3}{8}$

3. $\dfrac{5}{18} + \dfrac{7}{18}$

4. $\dfrac{-1}{24} + \dfrac{11}{24}$

5. $\dfrac{1}{10} + \dfrac{3}{10} + \dfrac{9}{10}$

6. $\dfrac{2}{15} + \dfrac{-4}{15} + \dfrac{8}{15}$

See Example 20

7. $\dfrac{10}{13} - \dfrac{2}{13}$

8. $\dfrac{-5}{9} - \dfrac{2}{9}$

9. $\dfrac{7}{10} - \dfrac{9}{10}$

10. $\dfrac{3}{11} - \dfrac{-5}{11}$

11. $\dfrac{7}{16} + \dfrac{5}{16} - \dfrac{3}{16}$

12. $\dfrac{11}{20} - \dfrac{9}{20} - \dfrac{7}{20}$

See
Example 21

13. $\dfrac{2}{x} + \dfrac{3}{x}$

14. $\dfrac{7}{a} + \dfrac{-8}{a}$

15. $\dfrac{x}{a+b} + \dfrac{2x-1}{a+b}$

16. $\dfrac{3t}{x-y} + \dfrac{2t+7}{x-y}$

17. $\dfrac{3u-5}{u(u+7)} + \dfrac{4u+1}{u(u+7)}$

18. $\dfrac{7m+5}{m(m-3)} + \dfrac{4m-8}{m(m-3)}$

19. $\dfrac{5x-10}{x(x+8)} + \dfrac{x^2-x+4}{x(x+8)}$

20. $\dfrac{3r+5}{(r+1)(r+2)} + \dfrac{r^2-2r-1}{(r+1)(r+2)}$

See
Examples
21 and 22

21. $\dfrac{8}{t} - \dfrac{3}{t}$

22. $\dfrac{5}{m} - \dfrac{11}{m}$

23. $\dfrac{3x}{x+y} - \dfrac{4x-3}{x+y}$

24. $\dfrac{7a-5}{a+b} - \dfrac{5a-4}{a+b}$

25. $\dfrac{7k-3}{k(k+2)} - \dfrac{4k+2}{k(k+2)}$

26. $\dfrac{8m-5}{(m+1)(m+7)} - \dfrac{3m^2-4m+3}{(m+1)(m+7)}$

27. $\dfrac{5t-7}{t(t-5)} - \dfrac{6t+2}{t(t-5)} - \dfrac{-t+4}{t(t-5)}$

28. $\dfrac{x-9}{(x+2)(x-5)} - \dfrac{2x+3}{(x+2)(x-5)} - \dfrac{-7x+1}{(x+2)(x-5)}$

Life science
application

29. If a family has seven children, the likelihoods for the number of boys are given by the following table.

Number of Boys	0	1	2	3	4	5	6	7
Likelihoods	$\dfrac{1}{2^7}$	$\dfrac{7}{2^7}$	$\dfrac{21}{2^7}$	$\dfrac{35}{2^7}$	$\dfrac{35}{2^7}$	$\dfrac{21}{2^7}$	$\dfrac{7}{2^7}$	$\dfrac{1}{2^7}$

Find the sum of all the likelihoods and simplify.

5.5   ADDITION AND SUBTRACTION OF FRACTIONS (UNLIKE DENOMINATORS)

If we wish to add or subtract fractions with unlike denominators, we must first rewrite the fractions so that they do have the same denominator. To do this, we look for the **least common denominator (LCD),** which is the smallest number that all the denominators will divide into evenly. (Do not confuse this with the GCF.)

For example, consider $\dfrac{5}{12} + \dfrac{7}{18}$. Let us look at the various fractions that are equal to these two. (Recall that we can multiply the top and bottom of a fraction by the same number, and the results are equal.)

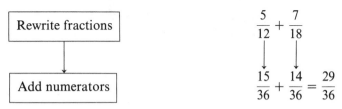

$$\frac{5}{12} = \frac{10}{24} = \boxed{\frac{15}{36}} = \frac{20}{48} = \frac{25}{60} = \cdots$$

$$\frac{7}{18} = \boxed{\frac{14}{36}} = \frac{21}{54} = \frac{28}{72} = \frac{35}{90} = \cdots$$

Notice that they both have an equivalent form with 36 as a denominator. Here, 36 is the LCD for 12 and 18 since it is the smallest number that both 12 and 18 divide into evenly. Now we can add the fractions.

Rewrite fractions	$\dfrac{5}{12} + \dfrac{7}{18}$
↓	↓
Add numerators	$\dfrac{15}{36} + \dfrac{14}{36} = \dfrac{29}{36}$

EXAMPLE 23   Add $\dfrac{5}{6} + \dfrac{1}{9}$.

SOLUTION   Here, the LCD for 6 and 9 is 18 (since 18 is the smallest number that both 6 and 9 will divide into evenly). We now rewrite the fractions with this LCD of 18. To do this, we multiply both fractions by 1 $\left(= \dfrac{3}{3} = \dfrac{2}{2} \right)$.

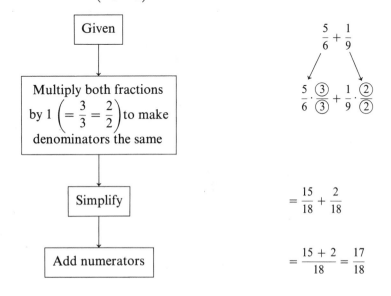

Given	$\dfrac{5}{6} + \dfrac{1}{9}$
Multiply both fractions by 1 $\left(= \dfrac{3}{3} = \dfrac{2}{2} \right)$ to make denominators the same	$\dfrac{5}{6} \cdot \dfrac{3}{3} + \dfrac{1}{9} \cdot \dfrac{2}{2}$
Simplify	$= \dfrac{15}{18} + \dfrac{2}{18}$
Add numerators	$= \dfrac{15 + 2}{18} = \dfrac{17}{18}$

We multiplied $\frac{5}{6}$ by $\frac{3}{3}$ since this gives us an equal fraction with a denominator of 18. Similarly, we multiplied $\frac{1}{9}$ by $\frac{2}{2}$ since this also gives a denominator of 18.

EXAMPLE 24   Subtract $\frac{11}{15} - \frac{4}{21}$.

SOLUTION   Sometimes, the LCD is not obvious. To find it, we factor the denominators into primes and line up the common primes.

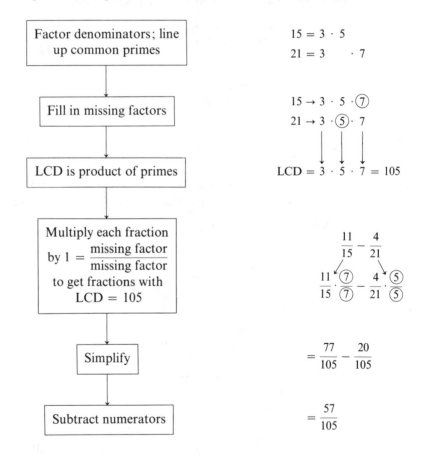

Notice that we use the missing factors to help us find the new equivalent fractions with LCD of 105.

EXAMPLE 25   Add $\dfrac{7}{8ab^2} + \dfrac{5}{6a^2b}$.

SOLUTION   These are algebraic fractions, but we use the same procedure that we used with the numeral fractions. We find the LCD by factoring the denominators.

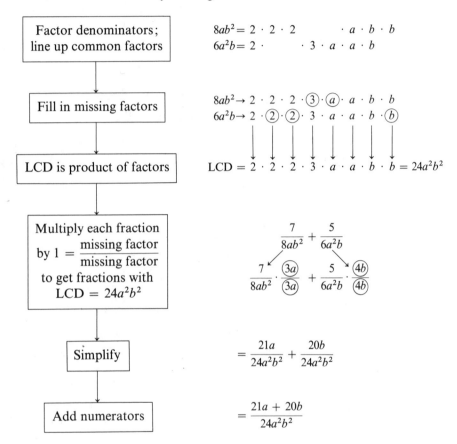

By factoring, we found the LCD as $24a^2b^2$. The circled missing factors, $3a$ and $4b$ $(= 2 \cdot 2 \cdot b)$, help us get the new fractions.

YES	NO
$\dfrac{7}{8ab^2} + \dfrac{5}{6a^2b} = \dfrac{21a + 20b}{24a^2b^2}$	$\dfrac{7}{8ab^2} + \dfrac{5}{6a^2b} = \dfrac{12}{24a^2b^2}$

After we find the LCD, we must also find the new numerators.

EXAMPLE 26   Add $\dfrac{5}{2x-8} + \dfrac{3x}{x^2-16}$.

SOLUTION   To find the LCD, we factor the denominators; then we use the missing factors to help us get the new fractions.

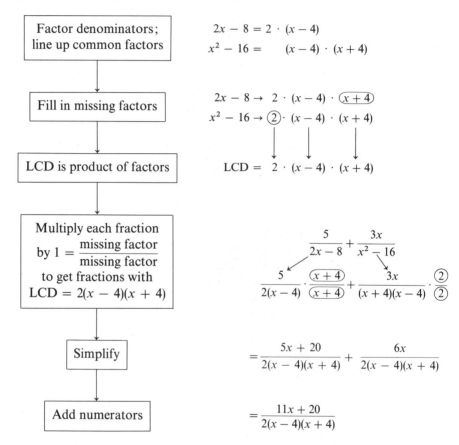

$$2x - 8 = 2 \cdot (x-4)$$
$$x^2 - 16 = (x-4) \cdot (x+4)$$

Factor denominators; line up common factors

$$2x - 8 \rightarrow 2 \cdot (x-4) \cdot \boxed{(x+4)}$$
$$x^2 - 16 \rightarrow \boxed{2} \cdot (x-4) \cdot (x+4)$$

Fill in missing factors

$$\text{LCD} = 2 \cdot (x-4) \cdot (x+4)$$

LCD is product of factors

Multiply each fraction by $1 = \dfrac{\text{missing factor}}{\text{missing factor}}$ to get fractions with LCD $= 2(x-4)(x+4)$

$$\dfrac{5}{2x-8} + \dfrac{3x}{x^2-16}$$

$$\dfrac{5}{2(x-4)} \cdot \dfrac{(x+4)}{(x+4)} + \dfrac{3x}{(x+4)(x-4)} \cdot \dfrac{2}{2}$$

Simplify

$$= \dfrac{5x+20}{2(x-4)(x+4)} + \dfrac{6x}{2(x-4)(x+4)}$$

Add numerators

$$= \dfrac{11x+20}{2(x-4)(x+4)}$$

EXAMPLE 27   Subtract $\dfrac{a}{a^2-36} - \dfrac{4}{a^2+5a-6}$.

SOLUTION   First, we change this to an addition problem by replacing the second fraction by its opposite:

$$\dfrac{a}{a^2-36} + \dfrac{-4}{a^2+5a-6}$$

Now we factor the denominators and find the LCD.

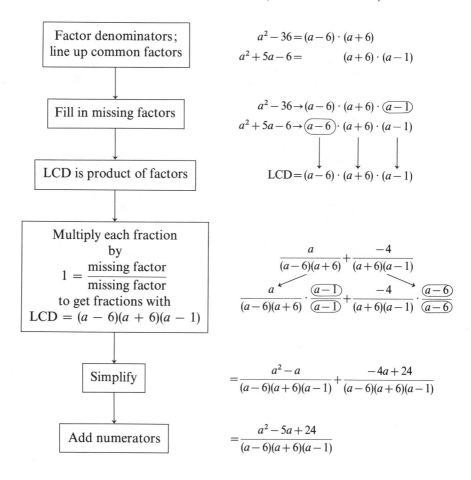

PROBLEM SET 5.5

Find the following sums or differences.

See
Example 23   **1.** $\dfrac{1}{3} + \dfrac{1}{2}$

2. $\dfrac{1}{7} - \dfrac{1}{4}$

3. $\dfrac{2}{5} - \dfrac{1}{10}$

4. $\dfrac{3}{8} + \dfrac{1}{2}$

5. $\dfrac{1}{6} - \dfrac{8}{9}$

6. $\dfrac{3}{10} + \dfrac{4}{15}$

See
Example 24   **7.** $\dfrac{11}{12} + \dfrac{5}{18}$

8. $\dfrac{13}{20} - \dfrac{7}{15}$

9. $\dfrac{7}{24} + \dfrac{3}{28}$

10. $\dfrac{6}{25} + \dfrac{8}{35}$

11. $\dfrac{13}{18} - \dfrac{5}{21}$

12. $\dfrac{7}{15} + \dfrac{3}{25}$

See Example 25   **13.** $\dfrac{1}{a} + \dfrac{1}{b}$

14. $\dfrac{1}{x} - \dfrac{1}{y}$

15. $\dfrac{3}{m} - \dfrac{m}{2}$

16. $\dfrac{5}{t} + \dfrac{t}{6}$

17. $\dfrac{2}{ab} - \dfrac{5}{3a}$

18. $\dfrac{5}{4x} + \dfrac{5}{xy}$

19. $\dfrac{8}{3x} + \dfrac{2}{x^2}$

20. $\dfrac{5}{4a} - \dfrac{3}{2a^2}$

21. $\dfrac{a}{2xy^2} - \dfrac{b}{4x^2y}$

22. $\dfrac{x}{10a^3b} + \dfrac{y}{5ab^2}$

23. $\dfrac{2}{3uv^2} + \dfrac{5}{6u^2w^2}$

24. $\dfrac{10a}{8m^2k^5} - \dfrac{b}{12n^3k^2}$

See Example 26   **25.** $\dfrac{5}{2x - 6} + \dfrac{7}{3x - 9}$

26. $\dfrac{2}{3x + 15} - \dfrac{5}{4x + 20}$

27. $\dfrac{x}{x^2 + 7x} - \dfrac{3}{3x + 21}$

28. $\dfrac{3a}{2a^2 + a} + \dfrac{5}{4a + 2}$

29. $\dfrac{8}{x + 2} + \dfrac{3}{x - 5}$

30. $\dfrac{6}{a - 7} - \dfrac{4}{a + 4}$

31. $\dfrac{7}{2k + 3} - \dfrac{1}{k + 2}$

32. $\dfrac{3}{3r - 7} + \dfrac{5}{r + 3}$

33. $\dfrac{5}{3x - 9} + \dfrac{2}{x^2 - 9}$

34. $\dfrac{3}{2x + 10} - \dfrac{4x}{x^2 - 25}$

35. $\dfrac{10}{a^2 - 4} - \dfrac{2}{a^2 + 2a}$

36. $\dfrac{7}{k^2 - 36} + \dfrac{5}{k^2 - 6k}$

See Example 27   **37.** $\dfrac{3}{a^2 - 9a + 20} + \dfrac{5}{a^2 - 11a + 30}$

38. $\dfrac{7}{x^2 + 3x + 2} - \dfrac{4}{x^2 + 4x + 3}$

39. $\dfrac{4}{x^2 - 1} - \dfrac{6}{x^2 - x - 2}$

40. $\dfrac{5}{k^2 - 12k + 35} + \dfrac{2}{k^2 - 25}$

41. $\dfrac{8}{m^2 + 12m + 36} - \dfrac{7}{m^2 + 5m - 6}$

42. $\dfrac{6}{u^2 - 4u + 4} + \dfrac{3}{u^2 - 7u + 10}$

43. $\dfrac{x - 5}{x^2 - 3x - 28} - \dfrac{x}{x^2 - 7x}$

44. $\dfrac{t + 7}{t^2 + 13t + 40} + \dfrac{t - 2}{t^2 + 5t - 24}$

45. $\dfrac{m}{m^2 + 8m + 7} + \dfrac{m + 1}{2m^2 + 3m + 1}$

46. $\dfrac{a - 1}{a^2 - 10a + 16} - \dfrac{a + 1}{3a^2 - 26a + 16}$

Business   **47.** At price p, the demand for a certain product is $\dfrac{10,000}{p}$. At an increased
application

price, $p + h$, the demand is $\dfrac{10,000}{p + h}$. Write the difference of these demands
as a single fraction:

$$\frac{10,000}{p + h} - \frac{10,000}{p}$$

Electricity   **48.** If two resistors R_1 and R_2 are connected in parallel as shown, the combined
application   resistance R is given by

$$\frac{1}{R} = \frac{1}{R_1} + \frac{1}{R_2}$$

Find the sum of the fractions on the right side of the equation. Then invert
both sides of the equation to find R.

Optical   **49.** The *lensmaker's equation* is given by
application

$$\frac{1}{f} = (n - 1)\left(\frac{1}{R_1} - \frac{1}{R_2}\right)$$

where f is the focal length of the lens, n is the index of refraction of glass, and R_1 and R_2 are the radii of curvature of the lens as shown. Write the right side of the equation as a single fraction. Then invert both sides of the equation to find the focal length f.

5.6   COMPLEX FRACTIONS

A **complex fraction** is a rational expression that contains at least one fraction in its numerator or denominator, or both. As examples,

$$\frac{\dfrac{2}{x}}{\dfrac{3}{y}}, \quad \frac{\dfrac{1}{2} - \dfrac{5}{12}}{\dfrac{3}{4} + \dfrac{1}{3}}, \quad \text{and} \quad \frac{1 - \dfrac{1}{m}}{2 + \dfrac{1}{m^2}}$$

are complex fractions. Each contains fractions within the fraction. We want to simplify these to simpler fractions.

EXAMPLE 28   Simplify $\dfrac{\dfrac{2}{x}}{\dfrac{3}{y}}$.

SOLUTION   There are two ways to work this problem. The first way is to treat this as a division problem.

$$\boxed{\frac{A}{B} \text{ means } A \div B}$$

Invert and multiply

$$\frac{\dfrac{2}{x}}{\dfrac{3}{y}} = \frac{2}{x} \div \frac{3}{y}$$

$$= \frac{2}{x} \cdot \frac{y}{3} = \frac{2y}{3x}$$

Another way to work this problem is to multiply the fraction, top and bottom, by the LCD, xy, of the two denominators x and y.

$$\boxed{\text{Multiply by } \frac{\text{LCD}}{\text{LCD}} = \frac{xy}{xy}}$$

$$\frac{\dfrac{2}{x}}{\dfrac{3}{y}} = \boxed{\frac{xy}{xy}} \cdot \frac{\dfrac{2}{x}}{\dfrac{3}{y}}$$

$$\downarrow$$

$$\boxed{\text{Cancel and simplify}}$$

$$= \frac{\cancel{xy}\left(\dfrac{2}{\cancel{x}}\right)}{\cancel{xy}\left(\dfrac{3}{\cancel{y}}\right)} = \frac{2y}{3x}$$

Perhaps both methods used in Example 28 seem about the same, but the second method is better for more involved problems.

To simplify complex fractions:

1. Find the LCD of all the denominators within the fraction.

2. Multiply the complex fraction by $1 = \dfrac{\text{LCD}}{\text{LCD}}$.

3. Simplify.

EXAMPLE 29 Simplify $\dfrac{\dfrac{1}{2} - \dfrac{5}{12}}{\dfrac{3}{4} + \dfrac{1}{3}}$.

SOLUTION We use the LCD method. The denominators are 2, 12, 3, and 4. Their LCD is 12, so we multiply by $\dfrac{12}{12}$.

$$\boxed{\text{Multiply by } \frac{\text{LCD}}{\text{LCD}} = \frac{12}{12}}$$

$$\frac{\dfrac{1}{2} - \dfrac{5}{12}}{\dfrac{3}{4} + \dfrac{1}{3}} = \boxed{\frac{12}{12}} \cdot \frac{\dfrac{1}{2} - \dfrac{5}{12}}{\dfrac{3}{4} + \dfrac{1}{3}}$$

$$\downarrow$$

$$\boxed{\text{Distributive law}}$$

$$= \frac{12\left(\dfrac{1}{2} - \dfrac{5}{12}\right)}{12\left(\dfrac{3}{4} + \dfrac{1}{3}\right)} = \frac{6 - 5}{9 + 4}$$

$$\downarrow$$

$$\boxed{\text{Simplify}}$$

$$= \frac{1}{13}$$

EXAMPLE 30   Simplify $\dfrac{1 - \dfrac{1}{m}}{2 + \dfrac{1}{m^2}}$.

SOLUTION   The LCD for the two denominators is m^2. Thus, we multiply the fraction by $\dfrac{m^2}{m^2}$.

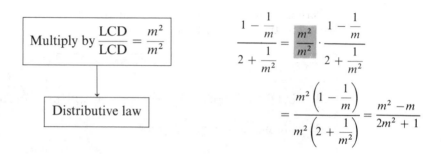

Notice how the LCD neatly clears away all the denominators.

PROBLEM SET 5.6

Simplify the following complex fractions.

See
Example 28

1. $\dfrac{\dfrac{2}{3}}{\dfrac{5}{7}}$

2. $\dfrac{\dfrac{5}{9}}{\dfrac{7}{11}}$

3. $\dfrac{\dfrac{3}{5}}{\dfrac{7}{10}}$

4. $\dfrac{\dfrac{2}{7}}{\dfrac{5}{14}}$

5. $\dfrac{\dfrac{3}{4}}{\dfrac{2}{3x}}$

6. $\dfrac{\dfrac{2}{5a}}{\dfrac{3}{6b}}$

7. $\dfrac{\dfrac{a}{b}}{\dfrac{c}{b^2}}$

8. $\dfrac{\dfrac{x}{y^2}}{\dfrac{t}{y^3}}$

See
Example 29

9. $\dfrac{\dfrac{1}{2} + \dfrac{1}{4}}{\dfrac{3}{8} - \dfrac{1}{4}}$

10. $\dfrac{\dfrac{1}{3} + \dfrac{1}{6}}{\dfrac{1}{12} + \dfrac{1}{6}}$

11. $\dfrac{1 + \dfrac{1}{3}}{2 + \dfrac{1}{4}}$

12. $\dfrac{3 - \dfrac{1}{15}}{2 + \dfrac{1}{10}}$

13. $\dfrac{\dfrac{2}{15} - \dfrac{3}{10}}{\dfrac{1}{5} + \dfrac{7}{30}}$

14. $\dfrac{\dfrac{7}{12} + \dfrac{5}{8}}{\dfrac{1}{4} - \dfrac{11}{24}}$

15. $\dfrac{2 - \dfrac{1}{12}}{\dfrac{11}{18} + \dfrac{13}{36}}$

16. $\dfrac{5 + \dfrac{1}{15}}{\dfrac{3}{4} - \dfrac{7}{20}}$

See Example 30

17. $\dfrac{\dfrac{1}{a} - \dfrac{1}{b}}{\dfrac{1}{a} + \dfrac{1}{b}}$

18. $\dfrac{\dfrac{1}{u} + \dfrac{1}{v}}{\dfrac{1}{u} - \dfrac{1}{v}}$

19. $\dfrac{2 - \dfrac{1}{x}}{3 + \dfrac{1}{y}}$

20. $\dfrac{4 + \dfrac{2}{m}}{5 - \dfrac{3}{n}}$

21. $\dfrac{6 + \dfrac{1}{a}}{7 - \dfrac{1}{a^2}}$

22. $\dfrac{4 - \dfrac{3}{t^2}}{7 - \dfrac{4}{t^3}}$

23. $\dfrac{\dfrac{1}{ab} - \dfrac{2}{a}}{2 + \dfrac{3}{b}}$

24. $\dfrac{\dfrac{2}{u^2 v} + \dfrac{5}{v}}{\dfrac{4}{u} - 1}$

Business application

25. The *elasticity* of a product's demand is given by

$$e = -\dfrac{\dfrac{\Delta Q}{Q}}{\dfrac{\Delta P}{P}}$$

where Q is the demand, ΔQ is the change in demand, P is the price, and ΔP is the change in the price. Write this as a simple fraction.

5.7   EQUATIONS WITH FRACTIONS

In Chapter 2 we studied linear equations that did not contain fractions. In this section we study equations with fractions, or **fractional equations.** As examples,

$$\frac{x}{2} - 1 = \frac{x}{3} \quad \text{and} \quad \frac{6}{x-3} + \frac{3}{2} = \frac{9}{x-3}$$

are fractional equations.

To solve these equations we first *multiply both sides by the LCD of all the denominators.* This clears away all the fractions.

EXAMPLE 31   Solve $\dfrac{x}{2} - 1 = \dfrac{x+1}{3}$.

SOLUTION   We clear away the fractions by multiplying both sides of the equation by the LCD, which is 6.

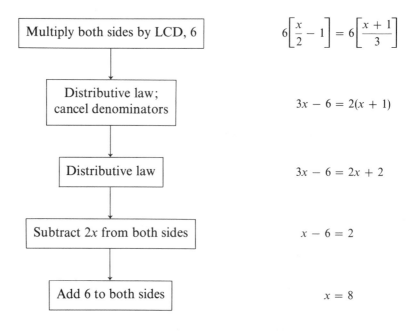

$$6\left[\frac{x}{2} - 1\right] = 6\left[\frac{x+1}{3}\right]$$

$$3x - 6 = 2(x + 1)$$

$$3x - 6 = 2x + 2$$

$$x - 6 = 2$$

$$x = 8$$

Let us check this by substituting $x = 8$ into the original equation.

$$Left\ side = \frac{8}{2} - 1 = 4 - 1 = 3$$

$$Right\ side = \frac{8+1}{3} = \frac{9}{3} = 3 \qquad Checks.$$

This example shows us the general procedure for solving these equations.

To solve fractional equations:

1. Determine the LCD of all the denominators.
2. Multiply both sides of the equation by this LCD. (This should clear away all the denominators.)
3. Solve the resulting equation. (Review Chapter 2.)
4. Be sure to check the solution in the original equation.

EXAMPLE 32   Solve $\dfrac{15}{a} + \dfrac{9a + 5}{a^2} = \dfrac{25}{a}$.

SOLUTION   The LCD is a^2, so we multiply both sides of the equation by a^2.

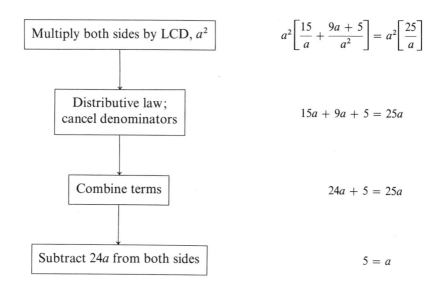

Multiply both sides by LCD, a^2	$a^2\left[\dfrac{15}{a} + \dfrac{9a + 5}{a^2}\right] = a^2\left[\dfrac{25}{a}\right]$
Distributive law; cancel denominators	$15a + 9a + 5 = 25a$
Combine terms	$24a + 5 = 25a$
Subtract 24a from both sides	$5 = a$

Let us check $a = 5$ in the original equation.

$$\textit{Left side} = \frac{15}{5} + \frac{9(5) + 5}{5^2} = \frac{15}{5} + \frac{50}{25} = 3 + 2 = 5$$

$$\textit{Right side} = \frac{25}{5} = 5 \quad \textit{Checks.}$$

EXAMPLE 33   Solve $\dfrac{3}{m - 1} - \dfrac{4}{m + 1} = \dfrac{6}{m^2 - 1}$.

SOLUTION   The LCD is $m^2 - 1 = (m + 1)(m - 1)$. To make the canceling easier, we multiply on on the left side of the equation by $(m - 1)(m + 1)$, and on the right by $m^2 - 1$. Since they are equal, this is valid.

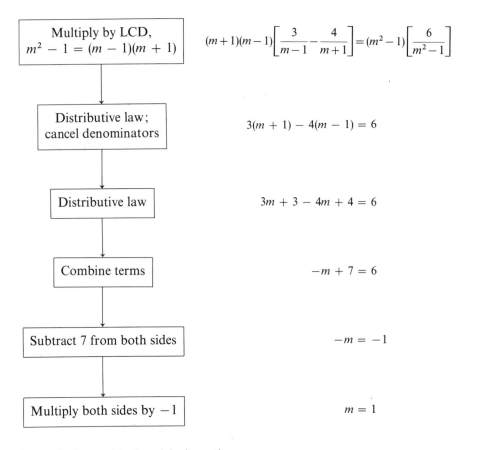

Multiply by LCD, $m^2 - 1 = (m-1)(m+1)$	$(m+1)(m-1)\left[\dfrac{3}{m-1} - \dfrac{4}{m+1}\right] = (m^2-1)\left[\dfrac{6}{m^2-1}\right]$
Distributive law; cancel denominators	$3(m+1) - 4(m-1) = 6$
Distributive law	$3m + 3 - 4m + 4 = 6$
Combine terms	$-m + 7 = 6$
Subtract 7 from both sides	$-m = -1$
Multiply both sides by -1	$m = 1$

Let us check $m = 1$ in the original equation.

$$Left\ side = \frac{3}{1-1} - \frac{4}{1+1} = \frac{3}{0} - \frac{4}{2} \ ??$$

$$Right\ side = \frac{6}{1^2 - 1} = \frac{6}{0} \ ??$$

The expressions $\dfrac{3}{0}$ and $\dfrac{6}{0}$ are meaningless. Since our procedure was correct, we have to conclude the $m = 1$ is *not* really a solution. Therefore, there is *no* solution.

Example 33 shows that we must always check, since it is possible to get a result that is not really a solution. The quickest way to spot this is as follows: *If the result produces a zero in any denominator, it is not a solution.*

PROBLEM SET 5.7

Solve the following equations. (Be sure to check all solutions—see Example 33.)

See Example 31

1. $\dfrac{x}{3} + 2 = \dfrac{3x - 3}{2}$

2. $a - \dfrac{a}{5} = \dfrac{7a + 5}{10}$

3. $\dfrac{4x + 1}{3} + 3 = \dfrac{16x - 2}{5}$

4. $-1 + \dfrac{m - 3}{4} = \dfrac{7 - m}{8}$

5. $\dfrac{t}{2} + \dfrac{t}{6} = \dfrac{t + 6}{3}$

6. $\dfrac{r}{5} + \dfrac{r}{4} = \dfrac{r}{10}$

7. $\dfrac{k}{2} + \dfrac{k}{4} = \dfrac{k + 4}{2}$

8. $\dfrac{u}{3} + 1 = \dfrac{u - 1}{2}$

See Example 32

9. $\dfrac{10}{p} + 2 = \dfrac{30}{p}$

10. $\dfrac{-8}{k} + 4 = \dfrac{-24}{k}$

11. $\dfrac{16}{u} + \dfrac{1}{2} = \dfrac{18}{u}$

12. $\dfrac{4}{m} - \dfrac{1}{3} = \dfrac{5}{m}$

13. $\dfrac{6}{x} - \dfrac{8}{x^2} = \dfrac{10}{x}$

14. $\dfrac{12}{a} - \dfrac{36}{a^2} = \dfrac{6}{a}$

15. $\dfrac{18}{t^2} + \dfrac{9}{t} = \dfrac{45}{t^2}$

16. $\dfrac{8}{m^2} - \dfrac{5}{m} = \dfrac{-12}{m^2}$

17. $\dfrac{12}{x - 3} - \dfrac{1}{2} = \dfrac{6}{x - 3}$

18. $\dfrac{3}{x + 7} + \dfrac{1}{4} = \dfrac{7}{x + 7}$

19. $\dfrac{12}{x - 1} + \dfrac{1}{2} = \dfrac{15}{x - 1}$

20. $\dfrac{3}{x + 4} + \dfrac{1}{3} = \dfrac{5}{x + 4}$

See Example 33

21. $\dfrac{2}{x + 1} + \dfrac{3}{x - 1} = \dfrac{11}{x + 1}$

22. $\dfrac{9}{a - 2} - \dfrac{14}{a + 2} = \dfrac{3}{a - 2}$

23. $\dfrac{3}{u + 4} + \dfrac{5}{u - 2} = \dfrac{2}{u + 4}$

24. $\dfrac{15}{r + 1} - \dfrac{8}{r - 2} = \dfrac{2}{r - 2}$

25. $\dfrac{6}{t - 1} - \dfrac{7}{t} = \dfrac{6}{t - 1}$

26. $\dfrac{10}{k - 5} + \dfrac{24}{k + 2} = \dfrac{20}{k - 5}$

27. $\dfrac{4}{m + 1} + \dfrac{5}{m - 1} = \dfrac{64}{m^2 - 1}$

28. $\dfrac{5}{x - 3} + \dfrac{2}{x + 3} = \dfrac{65}{x^2 - 9}$

29. $\dfrac{7}{c - 2} - \dfrac{10}{c + 3} = \dfrac{23}{c^2 + c - 6}$

30. $\dfrac{4}{w-3} + \dfrac{2}{w+5} = \dfrac{44}{w^2 + 2w - 15}$

Optical application **31.** For a thin lens, we have the formula

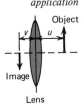

Object

Image

Lens

$$\frac{1}{f} = \frac{1}{u} + \frac{1}{v}$$

where f is the focal length, u is the distance to the object, and v is the distance to the image. Solve this equation for v if $f = 50$ millimeters and $u = 300$ millimeters.

Health application **32.** One formula for children's dosage of medicine is

$$C = \frac{A}{A + 12} D$$

where C is the child's dosage, A is the child's age, and D is the adult dosage. Solve this equation for the age A if $C = 40$ milliliters and $D = 100$ milliliters.

Electricity application **33.** If three capacitors are put in series, the total capacitance C is given by

$C_1 \quad C_2 \quad C_3$

$$\frac{1}{C} = \frac{1}{C_1} + \frac{1}{C_2} + \frac{1}{C_3}$$

Solve this equation for C_3 if $C = 10$ farads, $C_1 = 20$ farads, and $C_2 = 60$ farads.

5.8   WORD PROBLEMS

We now consider some word problems that involve fractional equations. Specifically, we look at **translation problems, work problems,** and **motion problems.**

EXAMPLE 34   What number must be added to both the numerator and denominator of $\dfrac{4}{13}$ to get the fraction $\dfrac{1}{2}$?

SOLUTION   This is a *translation problem.* We let x be the unknown.

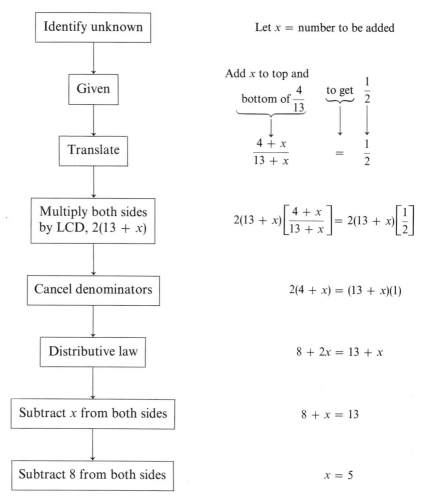

We can check this by adding 5 to the numerator and denominator:

$$\frac{4 + 5}{13 + 5} = \frac{9}{18} = \frac{1}{2}$$

EXAMPLE 35   Bob can mow a lawn in 3 hours, while Jim can mow the same lawn in 5 hours. How long will it take them working together?

SOLUTION   This is a *work problem*. We solve these by looking at one time period, such as 1 hour. Here, in 1 hour, Bob can mow $\frac{1}{3}$ of a lawn, while Jim can mow $\frac{1}{5}$ of a lawn. Also, if their combined time is x-hours, then in 1 hour, they can mow $\frac{1}{x}$ of a lawn together.

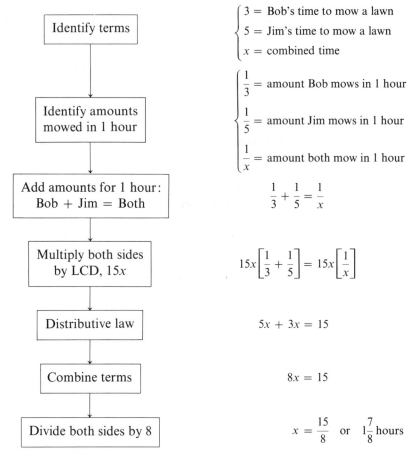

In general, suppose that two people (or machines, pipes, or whatever) can do jobs in times T_1 and T_2, and their combined time is x. Then the relationship is

$$\frac{1}{T_1} + \frac{1}{T_2} = \frac{1}{x}$$

(We subtract if the people or pipes are working against each other.)

EXAMPLE 36   In the same time, Carol drove 1000 miles while her mother drove only 600 miles. If Carol was averaging 20 miles per hour faster than her mother, what were their speeds?

SOLUTION   This is a *motion problem*. To work these problems, we use the following relations.

$$d = rt \qquad r = \frac{d}{t} \qquad t = \frac{d}{r}$$

where d is distance, r is rate (speed), and t is time. Also, when working these problems, it helps to make a d–r–t (distance-rate-time) table.

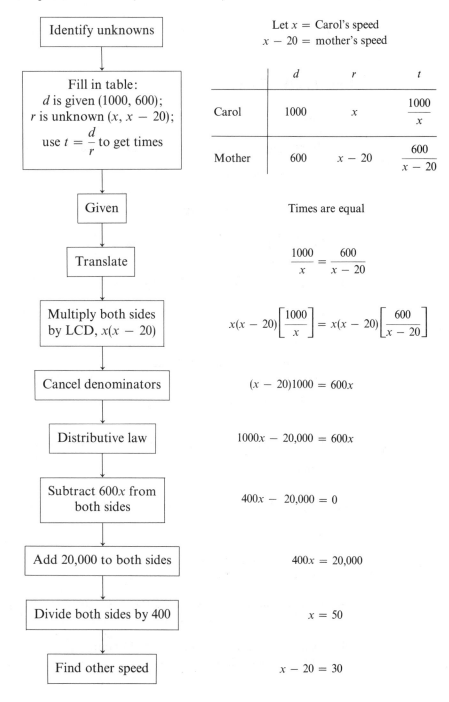

	d	r	t
Carol	1000	x	$\dfrac{1000}{x}$
Mother	600	$x - 20$	$\dfrac{600}{x - 20}$

Identify unknowns

Let $x =$ Carol's speed
$x - 20 =$ mother's speed

Fill in table:
d is given (1000, 600);
r is unknown $(x, x - 20)$;
use $t = \dfrac{d}{r}$ to get times

Given

Times are equal

Translate

$$\frac{1000}{x} = \frac{600}{x - 20}$$

Multiply both sides by LCD, $x(x - 20)$

$$x(x - 20)\left[\frac{1000}{x}\right] = x(x - 20)\left[\frac{600}{x - 20}\right]$$

Cancel denominators

$$(x - 20)1000 = 600x$$

Distributive law

$$1000x - 20{,}000 = 600x$$

Subtract $600x$ from both sides

$$400x - 20{,}000 = 0$$

Add 20,000 to both sides

$$400x = 20{,}000$$

Divide both sides by 400

$$x = 50$$

Find other speed

$$x - 20 = 30$$

Thus, Carol's speed was 50 and her mother's speed was 30. Notice that we use the fact that their times were the same. Also, we use the relation that $t = \dfrac{d}{r}$ to find the times in terms of the distances and rates.

PROBLEM SET 5.8

Solve the following word problems.

See
Example 34

1. What number must be added to both the numerator and denominator of $\dfrac{5}{17}$ to give the fraction $\dfrac{2}{3}$?

2. What number must be subtracted from both the numerator and denominator of $\dfrac{8}{11}$ to give the fraction of $\dfrac{1}{4}$?

3. What number must be added to the denominator of $\dfrac{9}{17}$ to give the fraction $\dfrac{1}{3}$?

4. What number must be subtracted from the numerator of $\dfrac{13}{15}$ to give the fraction $\dfrac{2}{3}$?

5. If $\dfrac{2}{3}$ is added to the reciprocal of a number, the result is $\dfrac{5}{6}$. Find the number.

6. If $\dfrac{1}{6}$ is subtracted from the reciprocal of a number, the result is $\dfrac{1}{30}$. Find the number.

See
Example 35

7. Ann can do a job in 6 hours, and Judy can do it in 10 hours. How long will it take them working together?

8. Ron can clean the litter in a park in 4 hours, and Mary can clean it in 5 hours. How fast can they clean the park if they work together?

9. Diane can canvass a precinct in 20 hours, and Dave can canvass it in 30 hours. How long will it take them if they work together?

10. In a certain post office, Joe can sort a stack of letters in 20 minutes, and Charlie can sort the same stack in 40 minutes. How long will it take them if they work together?

11. Pipe *P* can fill a pool in 5 hours, and pipe *Q* can empty it in 10 hours. If both pipes are open, how long will it take them to fill up an empty pool? (*Hint:* Subtract amounts.)

12. Pipe *A* can fill a tub in 7 minutes, and pipe *B* can empty it in 8 minutes. If both pipes are open, how long will it take to fill up an empty tub? (*Hint:* Subtract amounts.)

See Example 36

13. In the same time, Eddie drove 800 miles while Jerry drove 600 miles. Eddie's speed was 10 miles per hour faster than Jerry's. What were their speeds?

14. In the same time, jet *A* flies 1800 miles, and jet *B* flies 1650 miles. If jet *A* flies at a rate that is 50 miles per hour faster than jet *B*, what are their speeds?

15. In the same time, Jennifer rode her bike 20 miles while Melissa rode her bike 15 miles. If Jennifer rode her bike 5 miles per hour faster than Melissa, what were their speeds?

16. In the same time, Michael rowed his boat 8 miles while Steve rowed his boat 12 miles. If Steve rowed $\frac{2}{3}$ mile per hour faster than Michael, what were their speeds?

17. Sue made a 3000-mile trip in 35 hours. She traveled part of it in a 50-mile-per-hour car, and the rest on a 100-mile-per-hour train. What distances did she travel at each speed? (*Hint:* Let *x* be the distance in the car. Complete the following table and use the fact that the total time is 35 hours.)

	d	r	t
car	x		
train			

18. Larry traveled 540 miles in 13 hours. He drove part of it at 30 miles per hour (through towns), and the rest of it at 45 miles per hour (through the country). What distances did he drive at each speed? (See the hint for Problem 17.)

CHAPTER 5 SUMMARY

Important Words and Phrases

adding fractions (5.4)
algebraic fractions (5.1)
common denominator (5.4)
complex fraction (5.6)
denominator (5.1)
dividing fractions (5.3)
domain (5.1)
fraction (5.1)
fractional equation (5.7)
least common denominator (LCD)
  (5.5)
like denominators (5.4)

motion problems (5.8)
multiplicative inverse (5.3)
multiplying fractions (5.2)
numerator (5.1)
rational expression (5.1)
rational number (5.1)
reciprocal (5.3)
reduced to simplest terms (5.1)
subtracting fractions (5.4)
translation problems (5.8)
unlike denominators (5.5)
work problems (5.8)

Important Properties

For real numbers or polynomials A, B, C, and D,

$$\frac{A}{A} = 1 \qquad (A \neq 0)$$

$$\frac{A}{B} \cdot \frac{C}{D} = \frac{A \cdot C}{B \cdot D} \qquad (B, D \neq 0)$$

$$\frac{A}{B} = \frac{A}{B} \cdot \frac{C}{C} = \frac{A \cdot C}{B \cdot C} \qquad (B, C \neq 0)$$

$$\frac{A}{B} = \frac{A \div C}{B \div C} \qquad (B, C \neq 0)$$

$$A \cdot \frac{1}{A} = 1 \qquad (A \neq 0)$$

$$\frac{A}{B} \div \frac{C}{D} = \frac{A}{B} \cdot \frac{D}{C} = \frac{A \cdot D}{B \cdot C} \qquad (B, C, D \neq 0)$$

$$\frac{A}{C} + \frac{B}{C} = \frac{A + B}{C} \qquad (C \neq 0)$$

$$\frac{A}{C} - \frac{B}{C} = \frac{A - B}{C} \qquad (C \neq 0)$$

$$\frac{1}{T_1} + \frac{1}{T_2} = \frac{1}{x}$$ where T_1 and T_2 are individual times for a job, and x is the combined time.

$$d = rt; r = \frac{d}{t}; t = \frac{d}{r}$$ where d is distance, r is rate (speed), and t is the time.

Important Procedures

To reduce a fraction:

1. Factor the numerator and denominator completely.
2. Cancel common factors and simplify.

To multiply algebraic fractions:

1. Factor all numerators and denominators completely.
2. Multiply across, cancel, and simplify.

To divide algebraic fractions:

1. Invert the divisor.
2. Multiply the resulting fractions.

To find the LCD and add fractions with unlike denominators:

1. Factor the denominators and line up common factors.
2. Fill in the missing factors.
3. The LCD is the product of the factors.
4. Multiply each fraction by $1 = \dfrac{\text{missing factor}}{\text{missing factor}}$ to get the fractions to have the same denominator, the LCD.
5. Add the numerators.

To simplify complex fractions:

1. Find the LCD of all the denominators within the complex fractions.
2. Multiply the fraction by $1 = \dfrac{\text{LCD}}{\text{LCD}}$ and simplify.

To solve fractional equations:

1. Find the LCD of all the denominators.
2. Multiply both sides of the equations by the LCD.
3. Solve.
4. Check.

CHAPTER 5 REVIEW EXERCISES

Reduce the following fractions to simplest terms.

1. $\dfrac{12}{28}$

2. $\dfrac{96}{80}$

3. $\dfrac{18r^3s^4}{12r^5s^2}$

4. $\dfrac{5t + 15}{t^2 + 3t}$

5. $\dfrac{x^2 - 25}{2x + 10}$

6. $\dfrac{k - 3}{9 - k^2}$

7. $\dfrac{a^2 - a - 20}{a^2 - 16}$

Give the domain for each of the following fractions.

8. $\dfrac{t}{5}$

9. $\dfrac{3}{a - 6}$

Give the reciprocal for each of the following fractions.

10. $\dfrac{3a^2}{7}$

11. $\dfrac{2x - 5}{3x + 7}$

Perform the following operations. Write the answer in simplest terms.

12. $\dfrac{2}{5} \cdot \dfrac{15}{16}$

13. $\dfrac{-7}{8} \cdot \dfrac{20}{21}$

14. $\dfrac{6a^2b}{7c^4d^2} \cdot \dfrac{35cd^3}{36a^3b^3}$

15. $\dfrac{t^3 + 5t^2}{t^2 - 2t} \cdot \dfrac{t^2 - 4}{t^4}$

16. $\dfrac{x^2 + 3x - 4}{x^2 - 16} \cdot \dfrac{x^2 + x - 20}{x^2 + 6x - 7}$

17. $\dfrac{10}{21} \div \dfrac{5}{14}$

18. $\dfrac{3a^2b^4}{7c^2d} \div \dfrac{9a^3b^3}{14cd^5}$

19. $\dfrac{a^2}{a^2 - 2a} \div \dfrac{a^5}{a^2 - 4}$

20. $\dfrac{t^2 - 8t + 15}{t^2 - 9t + 20} \div \dfrac{t^2 - 14t + 48}{t^2 - 12t + 32}$

21. $\dfrac{5}{17} + \dfrac{11}{17}$

22. $\dfrac{2}{11} - \dfrac{7}{11} - \dfrac{-6}{11}$

23. $\dfrac{3r + 5}{(r - 3)(r + 4)} + \dfrac{5r - 8}{(r - 3)(r + 4)}$

24. $\dfrac{8k - 3}{a + b} - \dfrac{7k - 5}{a + b}$

25. $\dfrac{7}{15} - \dfrac{3}{10}$

26. $\dfrac{11}{35} - \dfrac{3}{25}$

27. $\dfrac{7}{5a^2} + \dfrac{4}{10a}$

28. $\dfrac{3}{a + 5} - \dfrac{5}{a - 4}$

29. $\dfrac{4}{x^2 - 3x - 4} + \dfrac{2x}{x^2 - 7x + 12}$

30. $\dfrac{\dfrac{2}{9a}}{\dfrac{4}{27}}$

31. $\dfrac{\dfrac{1}{4} - \dfrac{7}{12}}{\dfrac{5}{6} - \dfrac{1}{3}}$

32. $\dfrac{\dfrac{1}{a} - \dfrac{1}{a^2}}{5 + \dfrac{3}{a}}$

Solve the following equations.

33. $\dfrac{7x + 1}{3} - \dfrac{x + 2}{2} = \dfrac{5x + 8}{6}$

34. $\dfrac{35}{x} + \dfrac{50}{x^2} = \dfrac{45}{x}$

35. $\dfrac{3}{m + 1} - \dfrac{1}{m - 2} = \dfrac{1}{m^2 - m - 2}$

Solve the following word problems.

36. What number must be added to both the numerator and denominator of $\dfrac{2}{11}$ to give the fraction $\dfrac{1}{2}$?

37. Pam can paint a room in 3 hours, and Rudy can paint the same room in 7 hours. How long will it take them if they work together?

38. In the same time, Eric drove 360 miles while Matthew drove 440 miles. Matthew drove 10 miles per hour faster than Eric. Find their speeds.

6
GRAPHING AND
LINEAR SYSTEMS

6.1   LINEAR EQUATIONS IN TWO VARIABLES

The equation $2 + 3 = 5$ is true.

The equation $2 + 5 = 6$ is false.

The equation $2 + x = 7$ depends on x to be true.

Now, the equation $2x + y = 12$ depends on *both* x and y to be true.

In Chapter 2 we dealt with linear equations in one variable, for example, $9x + 5 = 23$. In this chapter we deal with linear equations in two variables, such as $2x + y = 12$. The **solutions** to such an equation are all the pairs of x- and y-values that make the equation true.

Consider the equation $2x + y = 12$. There are many x-and-y pairs that make this equation true. For instance, by substituting we can see that

If $x = 1$ and $y = 10$, it is true; $[2(1) + 10 = 12]$

If $x = 2$ and $y = 8$, it is true; $[2(2) + 8 = 12]$

If $x = 3$ and $y = 6$, it is true; $[2(3) + 6 = 12]$

If $x = 6$ and $y = 0$, it is true; $[2(6) + 0 = 12]$

If $x = -4$ and $y = 20$, it is true; $[2(-4) + 20 = 12]$

and so on

It is easier to write these solutions as **ordered pairs:** $(1, 10)$, $(2, 8)$, $(3, 6)$, $(6, 0)$, $(-4, 20)$, and so on. The first number in the pair is always the x-value, and the second is always the y-value. We call each pair an ordered pair since $(1, 10) \neq (10, 1)$. $[(1, 10)$ means that $x = 1$ and $y = 10$, while $(10, 1)$ means that $x = 10$ and $y = 1.]$

EXAMPLE 1   Consider $2x + 3y = 24$. Which of the following ordered pairs are solutions: $(6, 4)$, $(0, 8)$, $(4, 6)$, and $(15, -2)$?

SOLUTION   We substitute each (x, y) pair into the equation and determine if it is true or false.

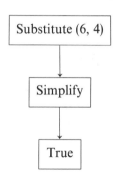

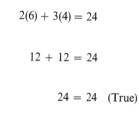

$2(6) + 3(4) = 24$

$12 + 12 = 24$

$24 = 24$ (True)

Thus, (6, 4) *is* a solution.

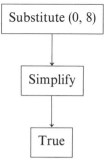

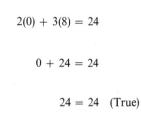

$2(0) + 3(8) = 24$

$0 + 24 = 24$

$24 = 24$ (True)

Thus, (0, 8) *is* a solution.

Substitute (4, 6)

Simplify

False

$2(4) + 3(6) = 24$

$8 + 18 = 24$

$26 = 24$ (False)

Thus, (4, 6) is *not* a solution.

Substitute (15, -2)

Simplify

True

$2(15) + 3(-2) = 24$

$30 + (-6) = 24$

$24 = 24$ (True)

Thus, (15, -2) *is* a solution.

Notice that (6, 4) *is* a solution, but (4, 6) is *not* a solution. This reminds us how important the order is.

Sometimes, we have a partial solution (one value, but not the other). *To find the missing value, substitute the known value into the equation and solve.*

EXAMPLE 2   For the equation $5x - 2y = 30$, complete the ordered pairs (4,) and $(-2,$).

SOLUTION   We find the missing y-values by substituting the given x-values into the equation and then solving.

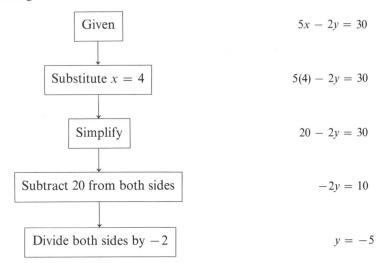

Therefore, the ordered pair is $(4, -5)$.

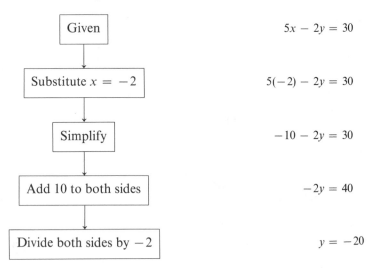

Therefore, the ordered pair is $(-2, -20)$.

EXAMPLE 3   For the equation $4x + 7y = 3$, complete the following ordered pairs: (, 0) and (, 1).

SOLUTION   We now substitute the y-values into the equation and solve for x.

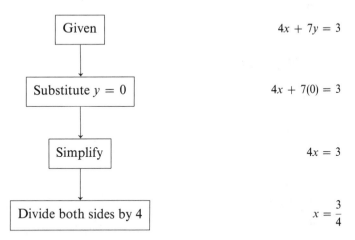

Therefore, the ordered pair is $\left(\dfrac{3}{4}, 0\right)$.

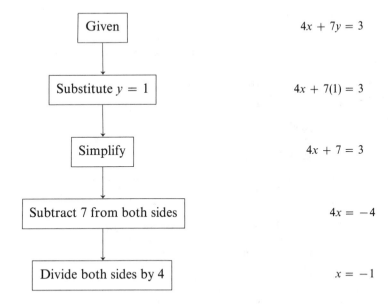

Therefore, the ordered pair is $(-1, 1)$.

EXAMPLE 4   For the equation $x = 6$, complete the ordered pairs: (, 8) and (, -4).

SOLUTION   The equation $x = 6$ probably looks strange since there is no y-term. This simply means that no matter what y is, x is *always* 6. Thus, the ordered pairs are (6, 8) and (6, −4).

Similarly, an equation such as $y = 2$ always has for its solutions a y-value of 2. For example, (3, 2), (−7, 2), (0, 2), and ($\frac{1}{2}$, 2) are all solutions to $y = 2$. (The second value is always 2.)

PROBLEM SET 6.1

For each equation, state which of the given ordered pairs is (are) solutions.

See Example 1

1. $x + y = 10$ (1, 9); (3, 8); (−4, 14)

2. $x − y = 6$ (10, 4); (−2, −8); (0, 6)

3. $2x + y = 10$ (1, 7); (0, 8); (3, 4)

4. $6x + y = 2$ (0, 2); (1, −1); (−1, 8)

5. $5x − 2y = 8$ (0, −4); (1, 2); (2, 1)

6. $4x − 5y = 12$ (3, 0); (1, −2); (5, −2)

7. $x − 7y = 10$ (3, −1); (17, 1); (10, 0)

8. $3x − 4y = 3$ (0, 1); (1, 0); (4, 3)

For each equation, complete the given ordered pairs.

See Example 2

9. $x + y = 10$ (4,   ) and (13,   )

10. $x − y = 12$ (20,   ) and (5,   )

11. $2x + y = 6$ (0,   ) and (6,   )

12. $3x + 2y = 8$ (2,   ) and (−4,   )

13. $5x − y = 3$ (1,   ) and (0,   )

14. $x − 3y = 12$ (0,   ) and (9,   )

See Example 3

15. $4x + 3y = 24$ (, 0) and (, 4)

16. $x − 2y = 18$ (, 1) and (, −2)

17. $3x − y = 21$ (, −6) and (, 15)

18. $−4x + 5y = 40$ (, 0) and (, 4)

19. $8x − y = 0$ (, −8) and (, 16)

20. $3x + 5y = 0$ (, 3) and (, −6)

See
Example 4 **21.** $x = 5$ (, 7) and (, −3)

22. $x = -6$ (, 0) and (, 2)

23. $y = 10$ (2,) and (−8,)

24. $y = -9$ (−5,) and (0,)

For each equation, complete the table.

See
Examples
2, 3, and 4 **25.** $3x - 4y = 12$

x	0		−4	
y		0		3

26. $7x - 2y = 14$

x	0		−2	
y		0		14

27. $x - 6y = 12$

x	0			
y		0	3	2

28. $-2x + 3y = 12$

x	0		3	9
y		0		

29. $x = 4$

x				
y	0	5	−7	12

30. $y = 8$

x	6	0	−15	9
y				

Physical
application **31.** The relation between hydrostatic pressure x and the water depth y is given by

$$33x + 15y = 495$$

Complete the ordered pairs (0,), (5,), and (, 0).

Health
application **32.** In a hospital, a certain patient needs 1 milligram of thiamine. He drinks prune juice x and orange juice y which satisfy the relation

$$0.05x + 0.08y = 1$$

Complete the ordered pairs (0,), (, 5), and (, 0).

Business   **33.** The relation between the price x of a certain item and its demand y is
applications   given by

$$1000x + y = 6000$$

Complete the ordered pairs (2,   ), (5,   ), and (, 2500).

34. A company buys a \$48,000 truck. As the truck ages and depreciates, the
relation between age x and its value y is given by

$$6000x + y = 48{,}000$$

Complete the ordered pairs (3,   ), (8,   ), and (, 18,000).

6.2   GRAPHING ORDERED PAIRS

To help us picture real numbers, we use a number line.

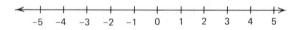

To help us picture ordered pairs of real numbers, we use the **rectangular coordinate system.** This is framed by a horizontal number line (called the **x-axis**), and a vertical number line (called the **y-axis**).

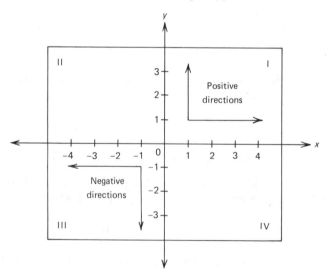

The plane is divided into four **quadrants,** labeled I, II, III, and IV, as shown. Every point of the plane is associated with an ordered pair of real numbers (called its **coordinates**). Notice that the *positive* directions are *up* and to the *right.* Also, the *negative* directions are *down* and to the *left.* The zero point at which the axes cross is called the **origin.** [This point has coordinates (0, 0).]

> For any ordered pair (x, y):
>
> **1.** x means the horizontal movement from 0 (right for +, left for −). This is called the **x-coordinate.**
> **2.** y means the vertical movement from 0 (up for +, down for −). This is called the **y-coordinate.**

EXAMPLE 5   Locate the points (2, 3), (4, −1), (3, 0), (−2, −3), (3, 2), and (−1, 3) on a graph.

SOLUTION   We determine what each pair means and then locate it on a graph.

(a) (2, 3) means 2 to the right, 3 up.

(b) (4, −1) means 4 to the right, 1 down.

(c) (3, 0) means 3 to the right, no units up or down.

(d) (−2, −3) means 2 to the left, 3 down.

(e) (3, 2) means 3 to the right, 2 up.

(f) (−1, 3) means 1 to the left, 3 up.

[Notice that (3, 2) and (2, 3) are *not* the same point on the graph.]

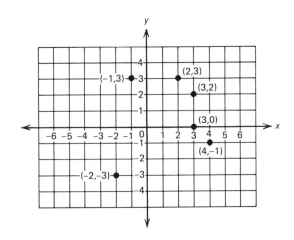

EXAMPLE 6   Give the coordinates of the points shown below.

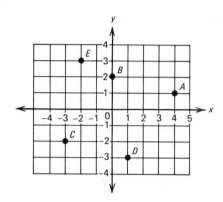

SOLUTION   For each point, we start at 0 and "walk" to the point (horizontally first, then vertically).

(a) *A* is 4 to the right, 1 up. Thus, $A = (4, 1)$.

(b) *B* is no units right or left, 2 up. Thus, $B = (0, 2)$.

(c) *C* is 3 to the left, 2 down. Thus, $C = (-3, -2)$.

(d) *D* is 1 to the right, 3 down. Thus, $D = (1, -3)$.

(e) *E* is 2 to the left, 3 up. Thus, $E = (-2, 3)$.

PROBLEM SET 6.2

Plot the following points.

See Example 5

1. (2, 3)	**2.** (4, 5)	**3.** (1, 4)
4. (5, 2)	**5.** (0, 3)	**6.** (0, −2)
7. (5, 0)	**8.** $(-\frac{1}{2}, 0)$	**9.** (−2, 3)
10. (−1, 5)	**11.** (−4, −2)	**12.** (−3, −3)
13. (2, −5)	**14.** (4, −3)	**15.** (1, −1)

Give the coordinates of the points on the graph.

See Example 6

16. Point *A*	**17.** Point *B*	**18.** Point *C*
19. Point *D*	**20.** Point *E*	**21.** Point *F*
22. Point *G*	**23.** Point *H*	**24.** Point *I*

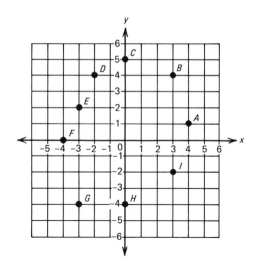

Estimate the coordinates of the points on the graph.

See
Example 6

25. Point *J*   **26.** Point *K*   **27.** Point *L*

28. Point *M*   **29.** Point *N*   **30.** Point *P*

31. Point *Q*   **32.** Point *R*   **33.** Point *S*

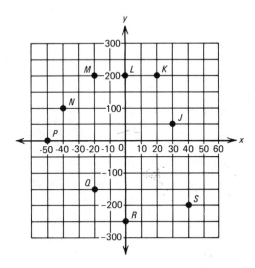

Business
application

34. The ABC YoYo Company has had its ups and downs lately. The graph indicates their stock prices over a certain 5-week period. The week's number is on the *x*-axis, and the stock price is on the *y*-axis. Give the coordinates of the points *A, B, C, D,* and *E*.

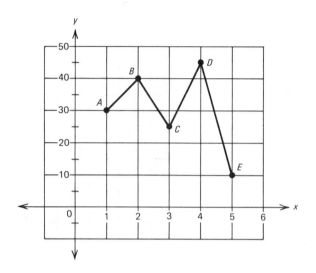

6.3   GRAPHING LINEAR EQUATIONS IN TWO VARIABLES

We are now ready to graph **linear equations with two variables.** In Section 6.1 we saw how to find some of the solutions to such linear equations. In Section 6.2 we saw how to plot such ordered pairs on a graph. We are now ready to put these skills together.

> To graph a linear equation with two variables:
>
> **1.** Find at least three solutions (ordered pairs) to the equation. [It is often helpful to make a table of the (x, y) pairs.]
> **2.** Plot these ordered pairs on a graph.
> **3.** Draw a straight line through these points.
> **4.** If the points do not form a straight line, you made a mistake somewhere; start over.

EXAMPLE 7   Graph the equation $x + y = 7$.

SOLUTION   This equation is so simple that we can very quickly see many solutions: $(0, 7)$, $(1, 6)$, $(2, 5)$, $(3, 4)$, $(4, 3)$, and so on. We are now ready to plot the points and form the line. We first list the solution points in a table.

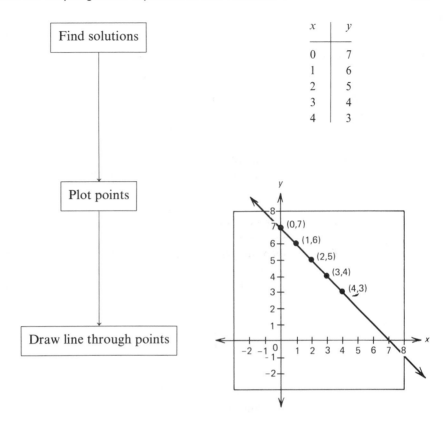

To help graph linear equations, we look for two special points: the **x-intercept,** the point where the line crosses the x-axis; the **y-intercept,** the point where the line crosses the y-axis. If we look at a graph, we see that the x-axis is where $y = 0$, and the y-axis is where $x = 0$. This gives the following rule.

RULE 1

1. To find the x-intercept, set $y = 0$ and solve the equation for x.
2. To find the y-intercept, set $x = 0$ and solve the equation for y.

EXAMPLE 8   Graph the equation $3x - 4y = 12$ by finding the x- and y-intercepts.

SOLUTION   We first find the x-intercept by setting $y = 0$.

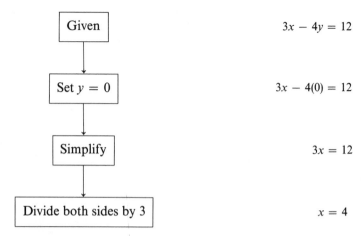

$$3x - 4y = 12$$

$$3x - 4(0) = 12$$

$$3x = 12$$

$$x = 4$$

Therefore, the x-intercept is $(4, 0)$. Now we find the y-intercept by setting $x = 0$.

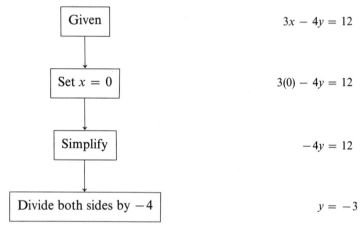

$$3x - 4y = 12$$

$$3(0) - 4y = 12$$

$$-4y = 12$$

$$y = -3$$

Therefore, the y-intercept is $(0, -3)$. Let us now find a third point. We pick a random x-value, $x = 2$, and solve for y.

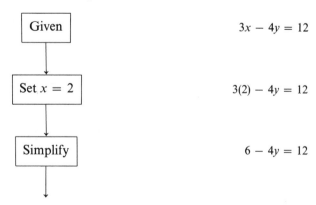

$$3x - 4y = 12$$

$$3(2) - 4y = 12$$

$$6 - 4y = 12$$

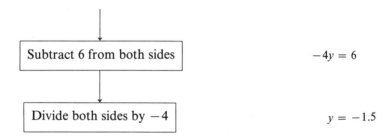

| Subtract 6 from both sides | $-4y = 6$ |

| Divide both sides by -4 | $y = -1.5$ |

Thus, our third point is $(2, -1.5)$. Now we make a table and graph these pairs.

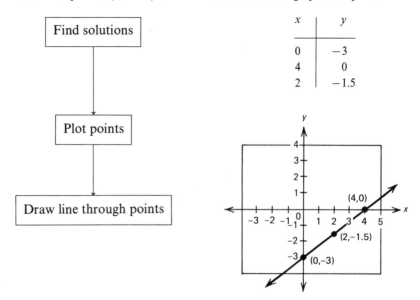

| Find solutions |

x	y
0	-3
4	0
2	-1.5

| Plot points |

| Draw line through points |

EXAMPLE 9   Graph $y = 3x - 2$.

SOLUTION   When the equation is in this form (with y alone), it is easiest to pick different x-values and solve for y. We let $x = 0, 1,$ and 2.

| Given | $y = 3x - 2$ |

| Set $x = 0$ | $y = 3(0) - 2$ |

| Simplify | $y = -2$ |

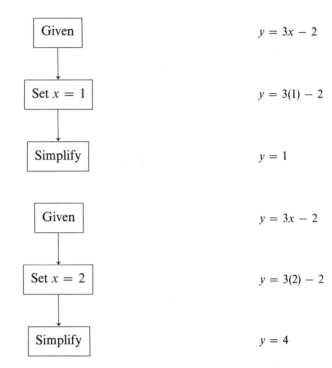

Given	$y = 3x - 2$
Set $x = 1$	$y = 3(1) - 2$
Simplify	$y = 1$
Given	$y = 3x - 2$
Set $x = 2$	$y = 3(2) - 2$
Simplify	$y = 4$

Therefore, our three points are $(0, -2)$, $(1, 1)$, and $(2, 4)$. We now graph these points.

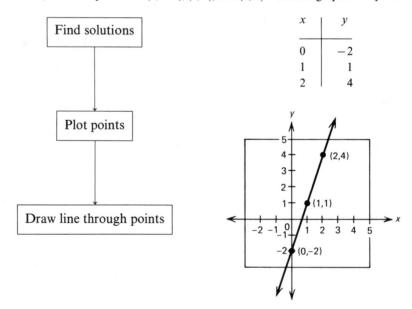

x	y
0	-2
1	1
2	4

EXAMPLE 10   Graph $x = 6$.

SOLUTION As we saw in Example 4, the equation $x = 6$ means that no matter what y is, x is always 6. For instance, $(6, 0)$, $(6, -3)$, and $(6, 4)$ are all solutions. (The first coordinate is always 6.)

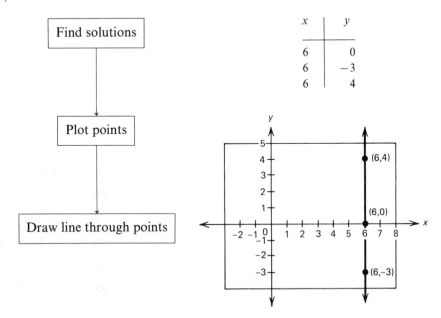

The equation $x = 6$ is a vertical line. In fact, the following table summarizes two special types of linear equations.

EQUATION	LINE	EXAMPLE
x = number	vertical	$x = 6$
y = number	horizontal	$y = -3$

PROBLEM SET 6.3

For each equation, complete the given ordered pairs; then use the points to graph the equation.

See Examples 2, 3, and 7

1. $x + y = 4$ $(0, \ \ ); (2, \ \ ); (\ \ , 0)$

2. $x - y = 5$ $(0, \ \ ); (7, \ \ ); (\ \ , 0)$

3. $2x + y = 10$ $(0, \ \ ); (3, \ \ ); (\ \ , 0)$

4. $4x - 2y = 8$ $(0, \ \ ); (1, \ \ ); (\ \ , 0)$

5. $x - 6y = 12$ $(0, \ ); (\ , 1); (\ , 0)$

6. $5x + 2y = 20$ $(0, \ ); (\ , 5); (\ , 0)$

7. $y = 2x + 1$ $(0, \ ); (1, \ ); (2, \ )$

8. $y = 8 - x$ $(0, \ ); (1, \ ); (2, \ )$

9. $x = 3$ $(\ , 4); (\ , -1); (\ , 0)$

10. $y = -2$ $(1, \ ); (4, \ ); (-3, \ )$

Graph the following linear equations.

See Example 8

11. $2x + 3y = 6$ 12. $3x - y = 3$

13. $4x - y = 8$ 14. $3x + 4y = 12$

15. $5x + 2y = 10$ 16. $x - 5y = 15$

17. $6x - y = 12$ 18. $4x + 2y = 16$

19. $3x + 5y = 15$ 20. $8x - 5y = 40$

21. $4x - 5y = 40$ 22. $2x + 7y = 14$

See Example 9

23. $y = 2x + 3$ 24. $y = x + 5$

25. $y = 4x$ 26. $y = -x$

27. $y = 3x + 2$ 28. $y = -2x + 3$

29. $y = 8 - x$ 30. $y = 12 - 3x$

See Example 10

31. $x = 7$ 32. $x = 3$

33. $x = 0$ 34. $x = -1$

35. $y = 5$ 36. $y = -1$

37. $y = 0$ 38. $y = \frac{1}{2}$

Business application

39. In producing a certain item, the relation between the cost y and the number of items produced x is given by

$$y = 10x + 1000$$

Graph this equation. (*Hint:* Scale the x-axis as 0, 100, 200, 300, and so on; scale the y-axis as 0, 1000, 2000, 3000, and so on.)

Health application

40. To get the proper amount of thiamine, one patient's intake of prune juice x and orange juice y must satisfy the formula

$$5x + 8y = 80$$

Graph this equation.

6.4   SOLVING LINEAR SYSTEMS BY GRAPHING

When we study two linear equations at the same time, we call this a **system of linear equations.** For example,

$$A: \quad x + y = 6$$
$$B: \quad x - y = 2$$

is a system of linear equations. We want a **simultaneous solution,** which is an ordered pair that makes *both* equations true at the same time.

EXAMPLE 11   Consider the system

$$A: \quad x + y = 6$$
$$B: \quad x - y = 2$$

(a) (5, 1) satisfies A (since $5 + 1 = 6$)
     but *not B* (since $5 - 1 \neq 2$)

(b) (7, 5) satisfies B (since $7 - 5 = 2$)
     but *not A* (since $7 + 5 \neq 6$)

(c) (8, 3) does *not* satisfy A (since $8 + 3 \neq 6$)
     and *not B* (since $8 - 3 \neq 2$)

(d) (4, 2) satisfies *both* A (since $4 + 2 = 6$)
     and B (since $4 - 2 = 2$)

Therefore, (4, 2) is our solution. It is the only point that satisfies *both* equations. We can see this by graphing the two equations.

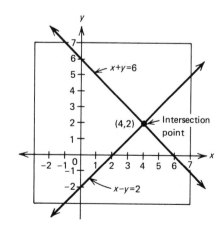

The solution to a system of linear equations is the intersection point (if any) of the two lines when graphed.

To solve a system of linear equations, A and B, by graphing:

1. Graph Equation A.
2. Graph Equation B using the same axes.
3. Find (or estimate) the coordinates of the intersection point (if there is one).
4. Check the solution in the original equations. It should make both of them true.

EXAMPLE 12   Solve the following system by graphing.

$$A: \quad x + 2y = 8$$
$$B: \quad -x + 3y = -3$$

SOLUTION   We graph lines A and B by finding points on each line.

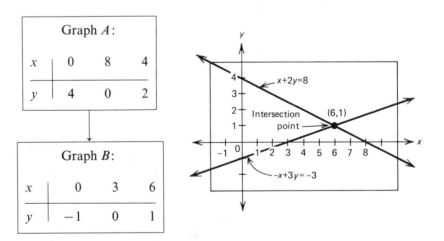

Graph A:

x	0	8	4
y	4	0	2

Graph B:

x	0	3	6
y	-1	0	1

The intersection of the lines appears to be at $(6, 1)$. Let us check this ordered pair in Equations A and B by substituting $x = 6$ and $y = 1$.

| Substitute (6, 1) into A and B |

$A:$ $6 + 2(1) = 8$ *Checks.*
$B:$ $-6 + 3(1) = -3$ *Checks.*

Thus, (6, 1) is the solution.

EXAMPLE 13   Solve the following system by graphing.

$$A:\ y = 6 - x$$
$$B:\ y = 2x - 3$$

SOLUTION   We graph both equations on the same axes.

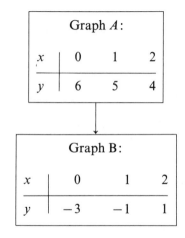

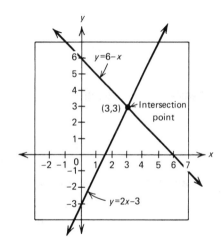

The intersection appears to be (3, 3). Let us check by substituting $x = 3$ and $y = 3$.

| Substitute (3, 3) into A and B |

$A:$ $3 = 6 - 3$ *Checks.*
$B:$ $3 = 2(3) - 3$ *Checks.*

Therefore, (3, 3) is the solution.

EXAMPLE 14   Solve the following system by graphing.

$$A:\ x + 2y = 2$$
$$B:\ 2x + 4y = 8$$

SOLUTION   We begin by graphing these equations on the same axes.

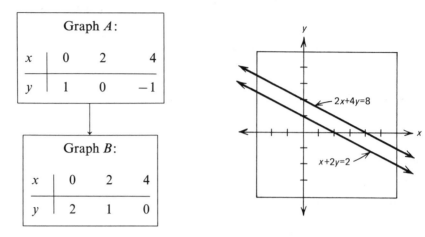

Graph *A*:			
x	0	2	4
y	1	0	−1

Graph *B*:			
x	0	2	4
y	2	1	0

There is no intersection since the lines are **parallel.** In this case we say that there is *no solution*.

EXAMPLE 15   Solve the following system by graphing.

$$A: \quad x - 3y = 3$$
$$B: \quad 2x - 6y = 6$$

SOLUTION   We graph both equations on the same axes.

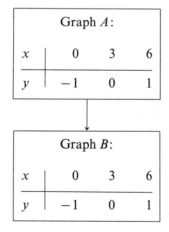

Graph *A*:			
x	0	3	6
y	−1	0	1

Graph *B*:			
x	0	3	6
y	−1	0	1

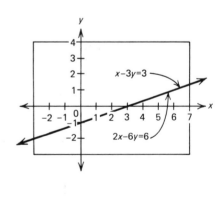

The two lines **coincide.** They are the *same* line. Therefore, the solution is the set of *all* points on either line.

We can summarize the three possibilities for the solutions to a system of two linear equations:

1. The lines intersect in exactly *one* point (ordered pair).
2. The lines are parallel with *no* solution.
3. The lines coincide (*same* line). Either line is the solution.

PROBLEM SET 6.4

For the following systems, test each of the given ordered pairs to decide if it is a solution to the system.

See
Example 11

1. A: $x + y = 12$
$\quad B$: $x - y = 6$
$\qquad (5, 7); (8, 2); (9, 3); (10, 2)$

2. A: $x + y = 9$
$\quad B$: $x - y = 13$
$\qquad (6, 3); (15, 2); (11, -2); (7, 2)$

3. A: $2x - y = 5$
$\quad B$: $3x + 5y = -12$
$\qquad (2, -1); (1, -3); (1, 2); (-1, 3)$

4. A: $2x + 3y = 16$
$\quad B$: $5x - 2y = 2$
$\qquad (5, 2); (4, 9); (2, 4); (-2, -6)$

5. A: $x + 4y = 4$
$\quad B$: $2x + 8y = 16$
$\qquad (0, 1); (8, 0); (4, 0); (0, 2)$

6. A: $x - 2y = 4$
$\quad B$: $-3x + 6y = -12$
$\qquad (0, -2); (4, 0); (2, -1); (6, 1)$

Solve the following systems of linear equations by graphing. [*Be careful;* some may be parallel lines (see Example 14), and some may be the same (coincident) line (see Example 15).]

See
Examples
11 and 12

7. A: $x + y = 10$
$\quad B$: $x - y = 6$

8. A: $x + y = 5$
$\quad B$: $x - y = 7$

9. A: $2x + y = 6$
$\quad B$: $x - y = -3$

10. A: $x - 2y = 8$
$\quad B$: $x + 3y = 3$

11. A: $2x + 3y = 6$
$\quad B$: $x + 2y = 5$

12. A: $x - 5y = 5$
$\quad B$: $3x + 5y = 15$

13. A: $2x - 3y = 12$
$\quad B$: $4x - 6y = 12$

14. A: $2x + y = 0$
$\quad B$: $-4x + 2y = 8$

15. A: $6x + 3y = 18$
   B: $x - y = -3$

16. A: $3x - y = 3$
   B: $-6x + 2y = 6$

See
Example 13 **17.** A: $y = 2x + 1$
   B: $y = 3x - 1$

18. A: $y = 2x + 8$
   B: $y = -x + 5$

19. A: $y = 4x + 2$
   B: $y = 5x + 2$

20. A: $y = 2x + 3$
   B: $y = 5$

21. A: $y = 2x + 1$
   B: $y = 2x + 2$

22. A: $y = 5x - 4$
   B: $y = x + 4$

23. A: $y = 4$
   B: $y = 3x - 2$

24. A: $x = 3$
   B: $y = 2x + 1$

25. A: $y = 3$
   B: $x = 2$

26. A: $y = -1$
   B: $x = 4$

27. A: $x = 6$
   B: $x = 2$

28. A: $y = 3$
   B: $y = -1$

Electricity
application
29. The circuit shown has currents x and y through resistors of 6 ohms and 3 ohms. Kirchhoff's and Ohm's laws lead to the following system:

$$A: \quad x - y = 0$$
$$B: \quad 6x + 3y = 18$$

Solve this system for the currents x and y by graphing.

18 *V*

6Ω 3Ω

6.5   SOLVING LINEAR SYSTEMS BY THE ELIMINATION METHOD

Solving systems of linear equations by graphing has its drawbacks:

1. It is time consuming to prepare a graph.

2. It is not always accurate, especially for solutions such as $\left(\dfrac{17}{23}, \dfrac{-14}{15}\right)$.

We now give another method for solving systems of linear equations. This method is algebraic, not graphic, and is called the **elimination method**. It involves eliminating a variable to reduce the system to one equation and one unknown.

First, we recall the addition property for equality: *If equal quantities are added to equal quantities, the results are also equal.* In symbols,

---

PROPERTY 1   If $A = B$ and $C = D$, then $A + C = B + D$.

We see how this is used in the examples that follow.

EXAMPLE 16   Solve the following system.

$$A: \quad x + y = 6$$

$$B: \quad x - y = 4$$

SOLUTION   If we add these two equations together, the y (from A) and the $-y$ (from B) cancel, leaving only x-terms.

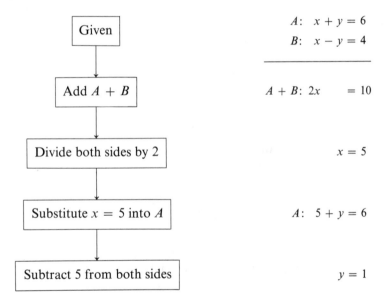

Given	$A: \quad x + y = 6$
	$B: \quad x - y = 4$
Add $A + B$	$A + B: 2x \qquad = 10$
Divide both sides by 2	$x = 5$
Substitute $x = 5$ into A	$A: \quad 5 + y = 6$
Subtract 5 from both sides	$y = 1$

Our solution is $x = 5$ and $y = 1$, or simply $(5, 1)$. We check this by substituting into Equation B:

$$B: \quad 5 - 1 = 4 \qquad Checks.$$

Notice that by adding Equations A and B together, we were able to eliminate y, and then solve a simpler equation, $2x = 10$. Once we had $x = 5$, we simply substituted this back into A to find y.

EXAMPLE 17   Solve the following system:

$$A: \quad x - y = 6$$
$$B: \quad 3x + 2y = -7$$

SOLUTION   Here, we do not have an obvious elimination as we did in Example 16. However, if we could make the $-y$ in Equation A into a $-2y$, it would cancel the $2y$ in Equation B upon adding the equations. We can do this by multiplying both sides of A by 2.

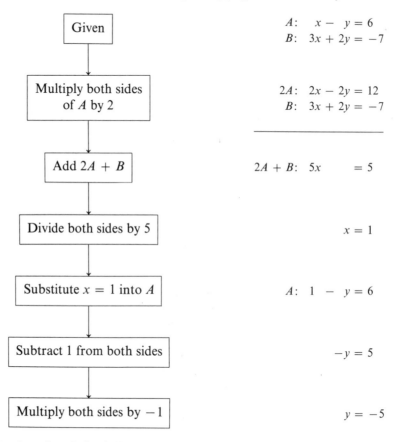

Given	$A:$	$x - y = 6$
	$B:$	$3x + 2y = -7$

Multiply both sides of A by 2	$2A:$	$2x - 2y = 12$
	$B:$	$3x + 2y = -7$

Add $2A + B$ $2A + B:$ $5x \quad\quad = 5$

Divide both sides by 5 $x = 1$

Substitute $x = 1$ into A $A:$ $1 - y = 6$

Subtract 1 from both sides $-y = 5$

Multiply both sides by -1 $y = -5$

Therefore, the solution is $(1, -5)$. Let us check this in Equation B:

$$B: \quad 3(1) + 2(-5) = 3 + (-10) = -7 \qquad \textit{Checks.}$$

EXAMPLE 18   Solve the following system.

$$A: \quad 2x - 3y = 2$$
$$B: \quad 5x + 2y = 24$$

SOLUTION Here it is not obvious what can be eliminated. Therefore, we simply choose the variable y (arbitrarily) and do what we have to do to eliminate it. If we multiply Equation A by 2, this will produce a $-6y$. Similarly, if we multiply Equation B by 3, this will produce a $6y$. Together, these terms cancel when we add.

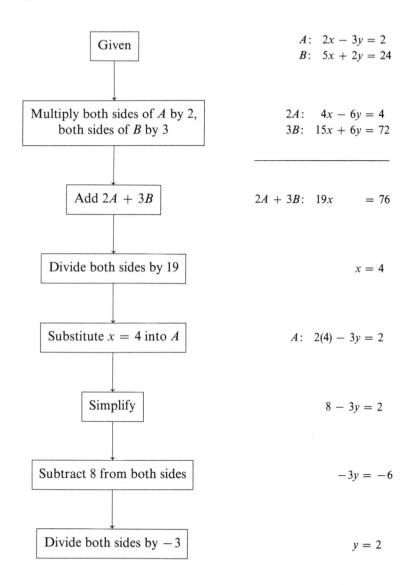

| Given | | A: $2x - 3y = 2$ |
| | | B: $5x + 2y = 24$ |

Multiply both sides of A by 2, both sides of B by 3
$$2A: \quad 4x - 6y = 4$$
$$3B: \quad 15x + 6y = 72$$

Add $2A + 3B$
$$2A + 3B: \quad 19x \qquad = 76$$

Divide both sides by 19
$$x = 4$$

Substitute $x = 4$ into A
$$A: \quad 2(4) - 3y = 2$$

Simplify
$$8 - 3y = 2$$

Subtract 8 from both sides
$$-3y = -6$$

Divide both sides by -3
$$y = 2$$

The solution is $(4, 2)$. We can check this in Equation B:

$$B: \quad 5(4) + 2(2) = 20 + 4 = 24 \qquad \textit{Checks.}$$

Before continuing, let us summarize the steps in the elimination method.

To solve a system of linear equations using the elimination method:

1. Write both equations in the form $ax + by = c$.

2. Choose a variable (x or y) to be eliminated.

3. Multiply each equation by a suitable number that makes the coefficients of the variable to be eliminated the opposites of each other (such as $5y$ and $-5y$).

4. Add the equations.

5. Solve for the remaining variable.

6. Substitute this value back into one of the original equations and solve for the other variable.

7. Check the solution.

Note: Step 4 may produce two unusual results:

(a) A *false* result, such as $0 = 4$, means that the lines are parallel, and there is no solution.

(b) A *true* result, such as $0 = 0$, means that the lines or equations are the same.

EXAMPLE 19   Solve the following system.

$$A: \quad 3x + 4y = 6$$

$$B: \quad 2x + 5y = 11$$

SOLUTION   For a change, let us eliminate x. If we multiply Equation A by -2, we get $-6x$; if we multiply Equation B by 3, we get $6x$. These cancel when we add.

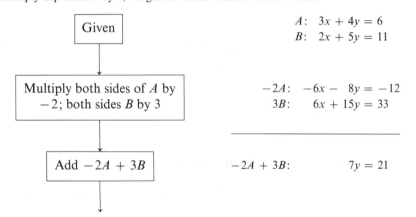

Given	$A: \quad 3x + 4y = 6$ $B: \quad 2x + 5y = 11$
Multiply both sides of A by -2; both sides B by 3	$-2A: \quad -6x - 8y = -12$ $3B: \quad 6x + 15y = 33$
Add $-2A + 3B$	$-2A + 3B: \qquad 7y = 21$

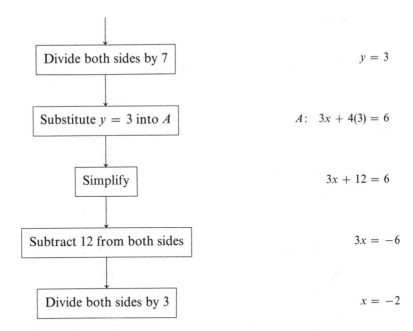

| Divide both sides by 7 | $y = 3$ |

| Substitute $y = 3$ into A | A: $3x + 4(3) = 6$ |

| Simplify | $3x + 12 = 6$ |

| Subtract 12 from both sides | $3x = -6$ |

| Divide both sides by 3 | $x = -2$ |

The solution is $(-2, 3)$.

EXAMPLE 20 Solve the following system.

$$A:\quad 2x - y = 6$$
$$B:\quad 4x - 2y = 16$$

SOLUTION If we multiply Equation A by -2, this will produce $-4x$, which cancels the $4x$ in Equation B when we add.

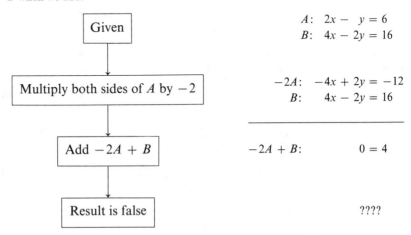

| Given | A: $2x - y = 6$
 B: $4x - 2y = 16$ |

| Multiply both sides of A by -2 | $-2A$: $-4x + 2y = -12$
 B: $4x - 2y = 16$ |

| Add $-2A + B$ | $-2A + B$: $0 = 4$ |

| Result is false | ???? |

The false result, $0 = 4$, indicates that there is *no* solution; graphically, the lines are parallel.

PROBLEM SET 6.5

Solve the following systems of linear equations using the elimination method. (*Be careful;* some may have no solution or may be the same line—see Example 20.)

See Example 16

1. A: $x + y = 10$
   B: $x - y = 2$

2. A: $x + y = 16$
   B: $x - y = -4$

3. A: $x + y = 7$
   B: $-x + y = 3$

4. A: $-x + y = 13$
   B: $x + y = -1$

5. A: $x - y = 6$
   B: $-x - y = 12$

6. A: $-x - y = 4$
   B: $x - y = 10$

7. A: $3x - y = 3$
   B: $4x + y = 11$

8. A: $-4x + y = 7$
   B: $6x - y = -5$

9. A: $x - 3y = 1$
   B: $-x + 5y = 7$

10. A: $x + 4y = 3$
   B: $-x + 2y = 15$

See Example 17

11. A: $x - y = 6$
   B: $2x + 3y = 17$

12. A: $x - y = 1$
   B: $4x + 5y = 49$

13. A: $x + y = -3$
   B: $3x - 4y = 19$

14. A: $x + y = 10$
   B: $6x - 5y = 38$

15. A: $3x - y = 17$
   B: $2x + 3y = 15$

16. A: $5x + y = -13$
   B: $3x - 2y = 0$

17. A: $x - 7y = 36$
   B: $2x - 3y = 17$

18. A: $-x + 2y = -2$
   B: $4x - 5y = 2$

19. A: $2x - y = -3$
   B: $6x + 5y = 15$

20. A: $x - 3y = 4$
   B: $2x - 6y = 3$

See Examples 18 and 19

21. A: $2x + 3y = 11$
   B: $3x - 4y = -9$

22. A: $5x - 3y = -20$
   B: $2x + 5y = 54$

23. A: $4x - 7y = 18$
   B: $2x - 3y = 8$

24. A: $-2x - 6y = -8$
   B: $3x - 5y = 12$

25. A: $4x + 7y = -18$
   B: $5x - 4y = 3$

26. A: $7x - 5x = 40$
   B: $-8x + 3y = -43$

27. A: $9x - 6y = -3$
   B: $-3x + 2y = 1$

28. A: $-10x - 3y = -54$
   B: $6x + 5y = 26$

29. A: $8x - 7y = 14$
   B: $9x + 5y = -10$

30. A: $5x - 2y = 12$
   B: $4x + 9y = -1$

31. A: $\quad 2x - 3y = 11$
$\quad\;\; B$: $\quad 3x + 2y = 23$

32. A: $\quad 4x - 3y = 12$
$\quad\;\; B$: $\quad 3x + 2y = 9$

33. A: $\quad 6x - 5y = 20$
$\quad\;\; B$: $\quad 4x + 3y = 26$

34. A: $\quad\;\; 7x - 2y = 11$
$\quad\;\; B$: $\quad -3x + 3y = 6$

35. A: $\quad 4x - 3y = -20$
$\quad\;\; B$: $\quad 5x + 2y = -2$

36. A: $\quad 8x - \;\; y = 2$
$\quad\;\; B$: $\quad 6x + 3y = -36$

37. An automobile company manufactures two cars: Squirrels and Beavers. The number of Squirrels x and Beavers y that they can produce is related to the amount of steel and labor available by the following system.

$$A: \quad 120x + \;\;\; 150y = \;\;\; 19{,}500$$
$$B: \quad 2400x + 3500y = 415{,}000$$

Solve this system for the number of each car produced.

38. A patient on a very restricted diet needs 72 grams of protein and 1.2 milligrams of thiamine. A dietitian uses slices of bread x and glasses of milk y and arrives at the following system.

$$A: \quad 6x + 8y = 120$$
$$B: \quad\;\; x + 4y = 36$$

Solve this system for the amounts of bread and milk.

6.6 SOLVING LINEAR SYSTEMS BY THE SUBSTITUTION METHOD

We have now studied the graphical and elimination methods for solving systems of linear equations. There is a third method that is sometimes more convenient: the **substitution method.** With this method, one of the variables is usually given alone in terms of the other; then we can substitute this variable into the other equation.

EXAMPLE 21 Solve the following system by substitution:

$$A: \quad 2x - 3y = -13$$
$$B: \quad\qquad y = 4x + 1$$

SOLUTION   Here Equation B gives y alone in terms of x. So, we take this expression for y and substitute it into Equation A in place of y. Then we solve the equation for x.

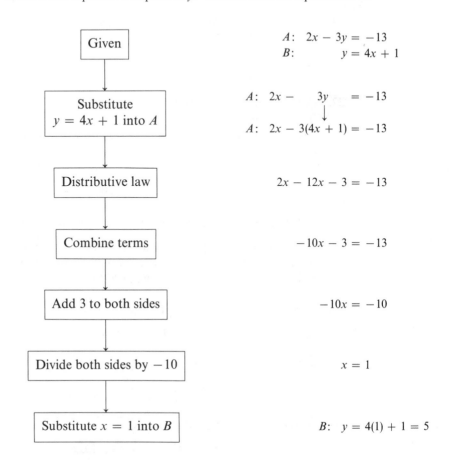

$$A:\ 2x - 3y = -13$$
$$B:\ \quad\quad y = 4x + 1$$

$$A:\ 2x - \quad 3y \quad = -13$$
$$\downarrow$$
$$A:\ 2x - 3(4x + 1) = -13$$

$$2x - 12x - 3 = -13$$

$$-10x - 3 = -13$$

$$-10x = -10$$

$$x = 1$$

$$B:\ y = 4(1) + 1 = 5$$

Therefore, the solution is $(1, 5)$.

EXAMPLE 22   Solve the following system by substitution.

$$A:\ 4x + 5y = 7$$
$$B:\ \quad\quad x = 2y - 8$$

SOLUTION   Here x is given in Equation B. We substitute this expression for x into Equation A.

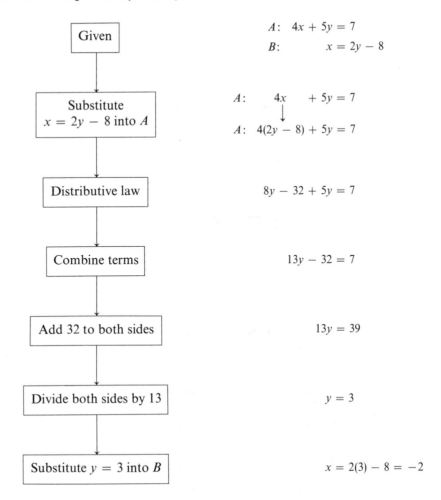

$$A:\quad 4x + 5y = 7$$
$$B:\qquad\quad x = 2y - 8$$

$$A:\qquad 4x \quad\; + 5y = 7$$
$$\downarrow$$
$$A:\quad 4(2y - 8) + 5y = 7$$

$$8y - 32 + 5y = 7$$

$$13y - 32 = 7$$

$$13y = 39$$

$$y = 3$$

$$x = 2(3) - 8 = -2$$

Therefore, the solution is $(-2, 3)$.

EXAMPLE 23   Solve the following system by substitution.

$$A:\quad 5x - 7y = 26$$
$$B:\quad\; x - 2y = 7$$

SOLUTION   Here neither equation has x or y alone, but Equation B can easily be solved for x.

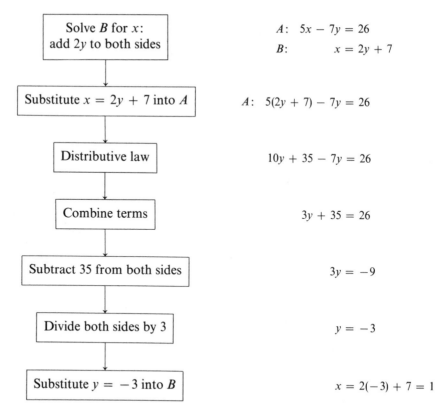

Solve B for x: add $2y$ to both sides	$A: \quad 5x - 7y = 26$ $B: \qquad x = 2y + 7$
Substitute $x = 2y + 7$ into A	$A: \quad 5(2y + 7) - 7y = 26$
Distributive law	$10y + 35 - 7y = 26$
Combine terms	$3y + 35 = 26$
Subtract 35 from both sides	$3y = -9$
Divide both sides by 3	$y = -3$
Substitute $y = -3$ into B	$x = 2(-3) + 7 = 1$

Therefore, the solution is $(1, -3)$.

EXAMPLE 24   Solve the following system by substitution.

$$A: \quad y = 2x - 14$$

$$B: \quad y = -3x + 6$$

SOLUTION   In this system, Equations A and B both express y in terms of x. We can substitute one of the expressions for y into the other equation. (We are actually setting the two y-expressions equal to each other.)

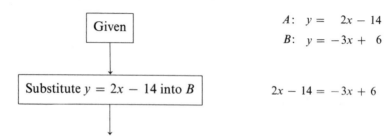

Given	$A: \quad y = \quad 2x - 14$ $B: \quad y = -3x + \ \ 6$
Substitute $y = 2x - 14$ into B	$2x - 14 = -3x + 6$

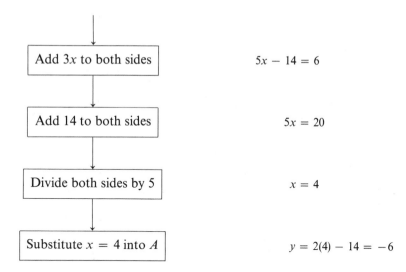

Add $3x$ to both sides	$5x - 14 = 6$
Add 14 to both sides	$5x = 20$
Divide both sides by 5	$x = 4$
Substitute $x = 4$ into A	$y = 2(4) - 14 = -6$

Thus, the solution is $(4, -6)$.

Just as in Section 6.5, a false result, such as $0 = 5$, indicates that there is no solution; also, a true result, such as $0 = 0$, indicates that the two lines are the same.

PROBLEM SET 6.6

Solve the following systems using the substitution method.

See
Example 21

1. A: $2x + 3y = 11$
        B: $\qquad y = 3x$

2. A: $5x - y = 2$
        B: $\qquad y = 4x$

3. A: $3x - 2y = -15$
        B: $\qquad y = -3x + 3$

4. A: $2x + 5y = 13$
        B: $\qquad y = x - 3$

5. A: $2x - 7y = -29$
        B: $\qquad y = 2x - 1$

6. A: $3x + 4y = -2$
        B: $\qquad y = 5x - 12$

7. A: $6x - 2y = 5$
        B: $\qquad y = 3x - 1$

8. A: $5x - y = 21$
        B: $\qquad y = -2x + 7$

See
Example 22

9. A: $4x - 3y = -5$
        B: $\qquad x = 2y$

10. A: $2x - 3y = 19$
         B: $\qquad x = -8y$

11. A: $2x + 5y = -39$
         B: $\qquad x = y + 5$

12. A: $4x + 7y = 21$
         B: $\qquad x = 2y - 6$

13. A: $6x - y = -19$
         B: $\qquad x = 2y + 6$

14. A: $-2x + 8y = -4$
         B: $\qquad x = 4y + 2$

See
Example 23

15. A: $2x + 3y = 4$
B: $x - y = 2$

16. A: $4x - 7y = 27$
B: $x - 2y = 7$

17. A: $3x - 2y = 7$
B: $x + y = -6$

18. A: $4x + 5y = 44$
B: $2x + y = 10$

19. A: $3x + 7y = 50$
B: $2x + y = 26$

20. A: $2x - 9y = 51$
B: $3x + y = 4$

21. A: $2x - 7y = 13$
B: $x - 3y = 13$

22. A: $3x - 6y = -10$
B: $-x + 2y = 5$

See
Example 24

23. A: $y = 2x + 1$
B: $y = 5x - 2$

24. A: $y = -x - 2$
B: $y = x + 14$

25. A: $y = 2x - 6$
B: $y = -x + 6$

26. A: $y = 3x - 9$
B: $y = 2x - 6$

27. A: $y = 4x + 5$
B: $y = 4x + 7$

28. A: $y = 4x - 1$
B: $y = 7x + 2$

29. A: $y = x - 4$
B: $y = 2x - 11$

30. A: $y = 5x - 29$
B: $y = 2x - 11$

*Consumer
application*

31. The annual cost y to operate a certain used car is given by $y = 800 + 0.10x$, where x is the number of miles driven per year. The annual cost to operate a certain new car is given by $y = 1400 + 0.07x$. Find the solution to the system of these two equations. (This is the point where the costs are the same.)

*Business
application*

32. The cost y to produce x-items of a certain product is given by $y = 100x + 30,000$. The revenue y from the sales of x-items is given by $y = 250x$. Solve the system of these two equations. (This is the *break-even point*, where the cost equals the revenue.)

*Population
application*

33. The enrollment at Beaver State College is 6000 and growing by 300 students per year. Its rival, Raccoon Community College, has 11,000 students, but is losing 200 students per year. If x is the number of years from now and y is enrollment, solve the following system to find when the enrollments will be the same.

$$A:\quad y = 6{,}000 + 300x$$

$$B:\quad y = 11{,}000 - 200x$$

6.7   WORD PROBLEMS

This final section deals with word problems. In many ways, these are similar to the word problems in Section 2.5, except that there are now two unknowns, x and y, and two equations.

To solve these word problems:

1. Read (and reread) the problem and identify the unknowns (label as *x* and *y*).
2. Translate the given information into two equations.
3. Solve this system.
4. Check the solutions.

EXAMPLE 25   The sum of two numbers is 44. Their difference is 6. Find the numbers.

SOLUTION   This problem is similar to the translation problems of Section 2.5. We begin by labeling the unknowns.

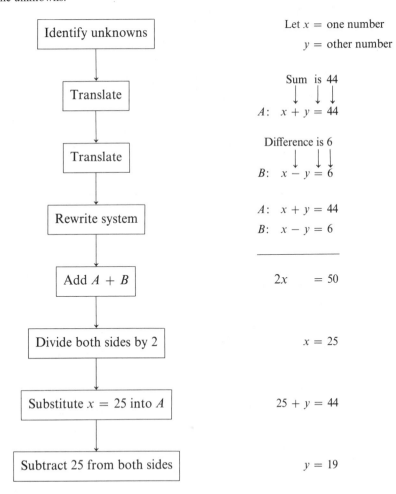

Identify unknowns

Let x = one number

y = other number

Translate

Sum is 44

A: $x + y = 44$

Translate

Difference is 6

B: $x - y = 6$

Rewrite system

A: $x + y = 44$
B: $x - y = 6$

Add $A + B$

$2x \quad = 50$

Divide both sides by 2

$x = 25$

Substitute $x = 25$ into A

$25 + y = 44$

Subtract 25 from both sides

$y = 19$

Therefore, the numbers are 25 and 19. We can check: their sum is 44, and their difference is 6.

EXAMPLE 26   The Smedleys invest $8000: some at 6% interest and the rest at 8% interest. Their total interest is $520. How much did they invest at each rate?

SOLUTION   For **interest problems** of this type, we look for two equations:

---

1. An investment equation:

investment I + investment II = total investment

2. An interest equation:

interest I + interest II = total interest

We start by letting x be the investment at 6% and y be the investment at 8%. Recall from arithmetic that 6% of x is $0.06x$, and 8% of y is $0.08y$. It is often helpful to fill in a table.

	INVESTMENT	RATE	INTEREST
I:	x	6%	$0.06x$
II:	y	8%	$0.08y$
Total:	8000		520

Equation A Equation B

Identify unknowns	Let x = investment at 6%
	y = investment at 8%

Identify interests	$0.06x$ = interest on 6% investment
	$0.08y$ = interest on 8% investment

Investment equation	investment I + investment II = total investment
Translate	x + y = 8,000

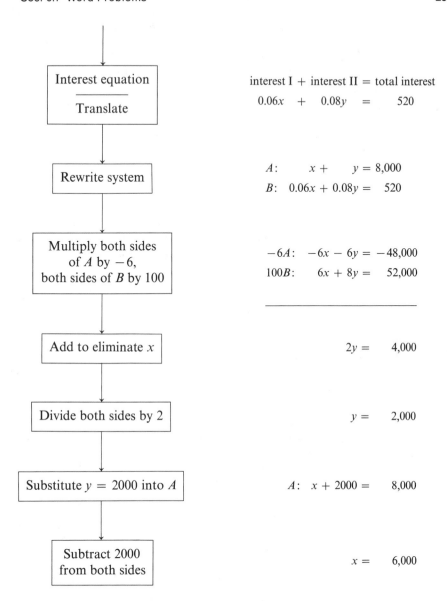

Interest equation ——— Translate	interest I + interest II = total interest
	0.06x + 0.08y = 520

A: x + y = 8,000
B: 0.06x + 0.08y = 520

Multiply both sides of A by −6, both sides of B by 100

−6A: −6x − 6y = −48,000
100B: 6x + 8y =   52,000

Add to eliminate x

2y = 4,000

Divide both sides by 2

y = 2,000

Substitute y = 2000 into A

A: x + 2000 = 8,000

Subtract 2000 from both sides

x = 6,000

Therefore, the solution is $6000 at 6% (= $360 interest) and $2000 at 8% (= $160 interest). This is a total of $520 interest.

EXAMPLE 27   How much 70% lean meat must a butcher mix with 90% lean meat to produce 100 pounds of 75% lean meat?

SOLUTION   For **mixture problems** of this type, we look for these two types of equations:

---

1. An amount equation (for the entire substance):

$$\text{amount I} + \text{amount II} = \text{total amount}$$

2. An ingredient equation (for the pure part):

$$\text{ingredient I} + \text{ingredient II} = \text{ingredient total}$$

Here we start by letting x be the amount of 70% lean meat and y be the amount of 90% lean meat. The amounts of the ingredient (pure lean) are 70% of x ($= 0.70x$) and 90% of y ($= 0.90y$). Altogether, this must be 75% of 100 pounds, or 75 pounds of pure lean. It is also helpful to fill in a table for mixture problems.

	AMOUNT (MEAT)	PERCENT LEAN	PURE LEAN
I:	x	70%	$0.70x$
II:	y	90%	$0.90y$
Total:	100	75%	$75\,[= (0.75) \cdot (100)]$

                            ↑   ↑

          Equation *A*   Equation *B*

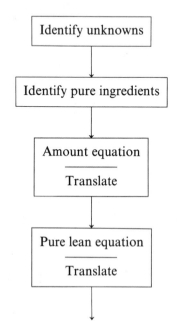

Let x = amount of 70% lean meat

y = amount of 90% lean meat

$0.70x$ = pure lean in 70% amount

$0.90y$ = pure lean in 90% amount

$$\begin{array}{ccccc} \text{amount I} & + & \text{amount II} & = & \text{total amount} \\ x & + & y & = & 100 \end{array}$$

$$\begin{array}{ccccc} \text{pure lean I} & + & \text{pure lean II} & = & \text{total pure lean} \\ 0.70x & + & 0.90y & = & 75 \end{array}$$

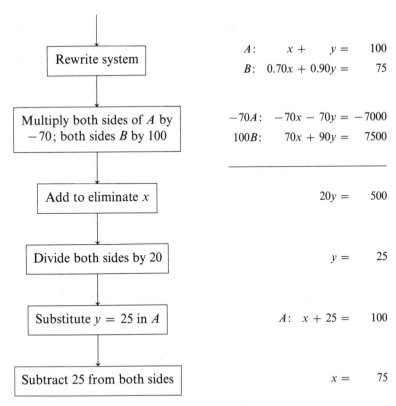

A: $x + y = 100$
B: $0.70x + 0.90y = 75$

$-70A$: $-70x - 70y = -7000$
$100B$: $70x + 90y = 7500$

$20y = 500$

$y = 25$

A: $x + 25 = 100$

$x = 75$

Thus, the butcher should mix 75 pounds of 70% lean meat (= 52.5 pounds lean) and 25 pounds of 90% lean meat (= 22.5 pounds lean). Together, this is 100 pounds of 75% lean meat (= 75 pounds lean).

EXAMPLE 28   A show sells $2 children's tickets and $3 adults' tickets. They sell a total of 500 tickets and receive $1320 in revenue. How many of each type of ticket were sold?

SOLUTION   We let x be the number of children's tickets sold, and let y be the number of adults' tickets sold. If there were x children's tickets sold, then they made $2x$ dollars ($2 per ticket). Similarly, if they sold y adults' tickets, then they made $3y$ dollars ($3 per ticket).

We have a *ticket equation* and a *money equation*. As in Examples 26 and 27, it is helpful to make a table.

	TICKETS	PRICE	VALUE
Children's:	x	2	$2x$
Adults':	y	3	$3y$
Total:	500		1320
	↑		↑
	Equation A		Equation B

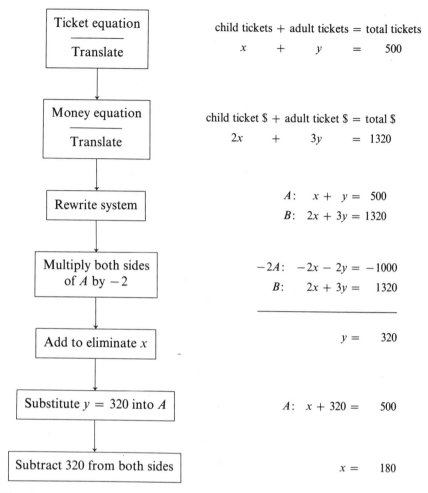

| Ticket equation | child tickets + adult tickets = total tickets |
| Translate | x + y = 500 |

| Money equation | child ticket \$ + adult ticket \$ = total \$ |
| Translate | $2x$ + $3y$ = 1320 |

Rewrite system

$$A:\quad x + y = 500$$
$$B:\quad 2x + 3y = 1320$$

Multiply both sides of A by -2

$$-2A:\quad -2x - 2y = -1000$$
$$B:\quad\ \ \ 2x + 3y =\ \ \ 1320$$

Add to eliminate x

$$y = 320$$

Substitute $y = 320$ into A

$$A:\quad x + 320 = 500$$

Subtract 320 from both sides

$$x = 180$$

Therefore, 180 children's tickets (= \$360) and 320 adults' tickets (= \$960) were sold for a total of \$1320.

PROBLEM SET 6.7

Work the following word problems.

See Example 25

1. The sum of two numbers is 47. Their difference is 11. Find the two numbers.

2. The sum of two numbers is 73. Their difference is 31. Find the two numbers.

3. One number is 6 times another number. Their sum is 56. Find the two numbers.

4. The larger of two numbers is 4 times the smaller number. Their difference is 21. Find the two numbers.

5. One number is 1 less than twice another number. Their sum is 17. Find the two numbers.

6. One number is 1 more than 4 times another number. Their sum is 21. Find the two numbers.

See Example 26

7. The Torres family invested $5000: some at 5% interest and the rest at 8% interest. Their total interest is $295. How much was invested at each amount?

8. Mr. LaRue invests $12,000: some at 6% interest and the rest at 10%. Together, he earns $760 interest. How much has he invested at each amount?

9. Ms. Morison invested $5000: some at 7%, the rest at 8%. Her total interest was $385. How much did she invest at each rate of interest?

10. Gooseville Junior College invests a $60,000 trust fund: some at 6% and the rest at 9%. Their total interest is $4860. How much was invested at each rate?

See Example 27

11. How many ounces of a 20% iodine solution and a 40% iodine solution must be mixed to produce 80 ounces of a 25% iodine solution?

12. How many milliliters of a 30% acid solution and a 50% acid solution must be mixed to get 300 milliliters of a 45% acid solution?

13. How many ounces of a 70% gold alloy and a 90% gold alloy must be mixed to get 40 ounces of an 85% gold alloy?

14. How many quarts of a 40% antifreeze solution must be mixed with a 90% antifreeze solution to produce 5 quarts of an 80% antifreeze solution?

See Example 28

15. A theater sells 1000 tickets to a show: some at $4, and the rest at $5. If the theater's revenue is $4400, how many of each are sold?

16. A bazaar sells 400 tickets: $0.50 for children, $1.50 for adults. If $300 is collected, how many of each are sold?

17. A confectioner has $3-per-pound candy and $4-per-pound candy. How much of each does she have to mix to get 40 pounds of $3.25-per-pound candy?

18. A patient in a hospital drank 14 ounces of juice: some prune juice, and the rest orange juice. An ounce of prune juice provides 5 milligrams of thiamine, and orange juice provides 8 milligrams of thiamine. If the patient got 100 milligrams of thiamine, how much of each juice did he drink?

CHAPTER 6 SUMMARY

Important Words and Phrases

coincident lines (6.4)
coordinates (6.2)
elimination method (6.5)
interest problems (6.7)
linear equations with two
   variables (6.3)
mixture problems (6.7)
ordered pair (6.1)
origin (6.2)
parallel lines (6.4)
rectangular coordinate system (6.2)

simultaneous solutions (6.4)
solutions (6.1)
substitution method (6.6)
system of linear equations (6.4)
x-axis (6.2)
x-coordinate (6.2)
x-intercept (6.3)
y-axis (6.2)
y-coordinate (6.2)
y-intercept (6.3)

Important Properties

For any ordered pair (x, y): x means horizontal movement (right for $+$, left for $-$), and y means vertical movement (up for $+$, down for $-$).

If $A = B$ and $C = D$, then $A + C = B + D$.

Important Procedures

To graph a linear equation with variables:
1. Find at least three solutions. (The x- and y-intercepts are convenient.)
2. Plot these solutions.
3. Draw a straight line through these points.

To solve a system of linear equations by graphing:
1. Graph the two equations on the same axes.
2. Estimate the intersection point (if any).

To solve a system of linear equations by elimination:
1. Write both equations in the form $ax + by = c$.
2. Choose the variable more easily eliminated.
3. Multiply each equation by a suitable number to make the coefficients of the variable to be eliminated the opposites of each other.
4. Add the equations.
5. Solve for the remaining variable.
6. Substitute this value into the original equation to find the other variable.

To solve a system of linear equations by substitution:
1. Write one of the variables in terms of the other.
2. Substitute this expression into the other equation.
3. Solve for the remaining variable.
4. Substitute this value into the original equation to find the other variable.

To solve the word problems of this chapter:
1. Read the problem and identify the unknowns by x and y.
2. Translate the information into two equations. (A table may be helpful.)
3. Solve the resulting system.

CHAPTER 6 REVIEW EXERCISES

For the equation $2x + 5y = 7$, which of the following are solutions?
1. $(0, 2); (1, 1); (6, -1); (-4, 3)$

For each equation, complete the given ordered pairs.

2. $2x + 3y = 12$ $(3, \ )$ and $(0, \ )$

3. $4x - 5y = 40$ $(\ , 0)$ and $(\ , 4)$

4. $x = 10$ $(\ , 3)$ and $(\ , -2)$

5. $2x - 7y = 14$

x	0		-7	
y		0		2

Plot the following points.

6. $(1, -3)$ **7.** $(-2, 0)$

Give the coordinates of the points on the graph.

8. Point A **9.** Point B

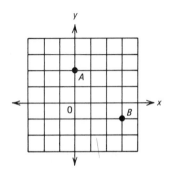

Graph the following linear equations.

10. $2x - 5y = 20$ **11.** $y = 3x - 2$

12. $x = -2$

Solve the following systems of linear equations: problems 13–15 by graphing, problems 16–18 by elimination, problems 19–21 by substitution.

13. A: $x - 2y = 8$ **14.** A: $y = 2x - 5$
         B: $x + 3y = 3$ B: $y = -x - 2$

15. A: $x + 3y = 6$ **16.** A: $2x - y = -1$
         B: $y = 1$ B: $5x + y = 15$

17. A: $x - y = -5$ **18.** A: $2x + 5y = 39$
         B: $3x + 2y = 0$ B: $7x - 3y = -7$

19. A: $5x - 2y = 25$ **20.** A: $2x + 7y = -13$
         B: $y = -2x + 1$ B: $x - 3y = 13$

21. A: $y = 5x - 1$
         B: $y = 5x + 2$

Solve the following word problems.

22. The sum of two numbers is 62. Their difference is 16. Find the two numbers.

23. The Conrads invest $2000: some at 7% and the rest at 8%. Their total interest is $147. How much did they invest at each rate?

24. A theater sells 300 tickets: some at $3 and the rest at $5 each. The total revenue is $1260. Find the number of each ticket sold.

7
ROOTS AND RADICALS

7.1   ROOTS AND RADICALS

In Chapter 3 we looked at the process of finding powers, such as **squaring.**

$$\textit{Squaring:} \quad 5^2 = \boxed{?}$$

Here, we had a base of 5, and we multiplied it by itself to get the result of 25. Now are going to reverse this process.

$$\textit{Square root:} \quad 25 = \boxed{?} \cdot \boxed{?} = \boxed{?}^2$$

Now we have 25, and we want to find its **square root:** What number squared gives 25 (or, what number times itself gives 25)? There are *two* answers: 5 and -5. This is true since $5 \cdot 5 = 25$ and $(-5)(-5) = 25$. Both are the square roots of 25.

EXAMPLE 1   The following are simple square roots.

(a) The square roots of 9 are 3 and -3.

(b) The square roots of 36 are 6 and -6.

(c) The square roots of 100 are 10 and -10.

(d) The square root of 0 is 0 (since $-0 = 0$).

When writing square roots, we use the symbol $\sqrt{}$. An expression such as $\sqrt{49}$ is called a **radical** and is *always the positive square root* of the number. The symbol $\sqrt{}$ is called the **radical sign,** and the number inside is called the **radicand.**

EXAMPLE 2   The following use the radical sign. Notice that the radical is always positive. If we want the negative square root, we put a negative sign in front of the radical.

(a) $\sqrt{64} = 8$

(b) $\sqrt{4} = 2$

(c) $\sqrt{121} = 11$

(d) $-\sqrt{49} = -7$

(e) $-\sqrt{81} = -9$

YES	NO
$\sqrt{64} = 8$	~~$\sqrt{64} = 8$ and -8~~

The *square roots* of 64 are 8 and -8, *but* $\sqrt{64}$ is only 8. The symbol $\sqrt{N}$ is always positive only.

EXAMPLE 3   Find $\sqrt{-9}$.

SOLUTION   What real number times itself gives -9? Is it 3? No, since $3 \cdot 3 = 9$. Is it -3? No, since $(-3)(-3) = 9$. Therefore, there is *no* real number that is $\sqrt{-9}$. If $-N$ is any negative number, there is *no* real number equal to $\sqrt{-N}$.

YES	NO
$-\sqrt{100} = -10$	~~$\sqrt{-100} = -10$~~

There is *no* real number $\sqrt{-100}$.

How about the square roots of numbers such as 11? Numbers such as 1, 4, 9, 16, 25, and so on are called **perfect squares.** Their square roots are easy to find. For numbers that are not perfect squares, we can use the table of **Square Roots and Cube Roots** (on the back endpaper) which gives three-place approximations.

EXAMPLE 4   The following use the Table of Squares and Square Roots on the back endpaper. (The symbol $\approx$ means "approximately equal to.")

(a) The square roots of 11 are approximately 3.317 and -3.317.

(b) The square roots of 37 are approximately 6.083 and -6.083.

(c) $\sqrt{73} \approx 8.544$

(d) $-\sqrt{46} \approx -6.782$

These square roots can also be found very simply with any hand calculator with a $\boxed{\sqrt{}}$ key. For example, let us find $\sqrt{21}$ and $\sqrt{6.92}$ on such a calculator.

PUNCH	DISPLAY	MEANING
$\boxed{C}$	0.	Clear
$\boxed{2}\,\boxed{1}$	21.	Enter 21
$\boxed{\sqrt{}}$	4.5825757	$\sqrt{21}$
$\boxed{C}$	0.	Clear
$\boxed{6}\,\boxed{\cdot}\,\boxed{9}\,\boxed{2}$	6.92	Enter 6.92
$\boxed{\sqrt{}}$	2.6305893	$\sqrt{6.92}$

Notice that the square roots given in Example 4 are approximations. The exact square roots of numbers such as 11, 37, 73, or 46 are not integers or fractions (or **rational numbers**). Rather, they are **irrational numbers** that must be approximated with a hand calculator or table of square roots.

EXAMPLE 5   The following table summarizes squares roots and radicals.

N NUMBER	$\sqrt{N}$ POSITIVE SQUARE ROOT	$-\sqrt{N}$ NEGATIVE SQUARE ROOT	NUMBER TYPE OF $\sqrt{N}$ (RATIONAL/IRRATIONAL)
36	6	-6	rational
81	9	-9	rational
0	0	0	rational
29	$\sqrt{29} \approx 5.385$	$-\sqrt{29} \approx -5.385$	irrational
93	$\sqrt{93} \approx 9.644$	$-\sqrt{93} \approx -9.644$	irrational
-4	no real number	no real number	—

In addition to square roots, we have other roots. The **cube root** of N answers the question: What number cubed gives N? Similarly, the **fourth root** of N answers the question: What number to the fourth power gives N? Higher roots are similarly defined. We write these with the radicals $\sqrt[3]{N}$, $\sqrt[4]{N}$, $\sqrt[5]{N}$, and so on. The digit in the radical sign is called the **index.** (We do not put a "2" in the square-root radical; it is understood.)

EXAMPLE 6   The following are examples of cube, fourth, and higher roots.

(a) $\sqrt[3]{64} = 4$ since $4^3 = 64$.

(b) $\sqrt[4]{81} = 3$ since $3^4 = 81$.

(c) $\sqrt[3]{125} = 5$ since $5^3 = 125$.

(d) $\sqrt[7]{128} = 2$ since $2^7 = 128$.

(e) $\sqrt[4]{-16}$ is not a real number since no real number to the fourth power gives -16.

PROBLEM SET 7.1

Find or approximate the following numbers. Use a hand calculator or table of square roots, if necessary. (*Be careful;* in some cases, no real number may exist—see Example 3.)

See Example 1

1. The square roots of 81.
2. The square roots of 4.
3. The square roots of 16.
4. The square roots of 1.
5. The square roots of -25.
6. The square roots of 49.
7. The square roots of 64.
8. The square roots of -36.
9. The square roots of 144.
10. The square roots of 400.

See Example 2

11. $\sqrt{25}$
12. $\sqrt{9}$
13. $\sqrt{0}$
14. $\sqrt{100}$
15. $-\sqrt{64}$
16. $-\sqrt{1}$
17. $\sqrt{-36}$
18. $-\sqrt{36}$
19. $-\sqrt{4}$
20. $-\sqrt{-16}$
21. $\sqrt{900}$
22. $\sqrt{2500}$

See Example 4

23. The square roots of 54.
24. The square roots of 79.
25. The square roots of 2.
26. The square roots of 5.
27. $\sqrt{18}$
28. $\sqrt{37}$
29. $-\sqrt{61}$
30. $-\sqrt{44}$
31. $\sqrt{-71}$
32. $-\sqrt{13}$

Complete the following table.

See Example 5

	N	$\sqrt{N}$	$-\sqrt{N}$	Number Type of $\sqrt{N}$ (Rational or Irrational?)
33.	100			
34.	4			
35.	12			
36.	38			
37.	-3			
38.	25			
39.		4		
40.		7		
41.			-3	
42.			-11	
43.		$\sqrt{6}$		
44.			$-\sqrt{15}$	

Find the indicated roots. [*Be careful;* some may not be real numbers—see Example 6(**e**).]

See Example 6

45. $\sqrt[3]{8}$ **46.** $\sqrt[3]{1}$ **47.** $\sqrt[3]{64}$

48. $\sqrt[3]{125}$ **49.** $\sqrt[3]{27}$ **50.** $\sqrt[3]{1}$

51. $\sqrt[4]{16}$ **52.** $\sqrt[4]{10{,}000}$ **53.** $\sqrt[5]{100{,}000}$

54. $\sqrt[4]{-81}$

Business application

55. The company finds that its sales S are given by

$$S = 1000\sqrt{A}$$

where A is their advertising budget. Find their sales S if
(a) $A = \$900$
(b) $A = \$10{,}000$
(c) $A = \$1600$

Engineering application

56. The maximal horizontal distance d that an antenna will project a signal is given by

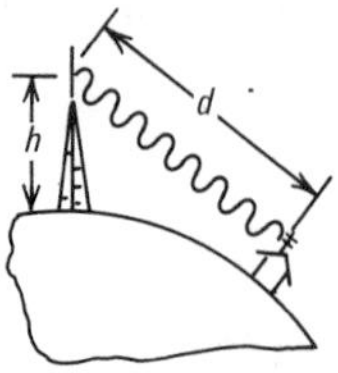

$$d = 1.22\sqrt{h}$$

where d is the distance in miles and h is the height of the antenna in feet. Compute the distance d for the following heights.

(a) $h = 400$ feet
(b) $h = 900$ feet
(c) $h = 225$ feet

Health   **57.** The pulse rate P in beats per minute is approximated by
application

$$P = \frac{600}{\sqrt{h}}$$

where h is a person's height in inches. Find P for the following heights. (Use a table or a calculator.)
(a) $h = 36$ inches
(b) $h = 60$ inches
(c) $h = 70$ inches

7.2   PROPERTIES OF RADICALS

Consider the following.

$$\sqrt{4}\sqrt{9} = 2 \cdot 3 = 6$$
$$\sqrt{4 \cdot 9} = \sqrt{36} = 6$$

Notice that $\sqrt{4 \cdot 9} = \sqrt{4}\sqrt{9}$. This property is true in general.

PROPERTY 1   For positive real numbers a and b,

$$\sqrt{ab} = \sqrt{a}\sqrt{b}$$

Although we consider only square roots in this section, Property 1 is also true for cube roots, fourth roots, and so on. We use this property to simplify radical expressions.

EXAMPLE 7   The following examples use Property 1.

(a) $\sqrt{3}\sqrt{5} = \sqrt{15}$
(b) $\sqrt{3}\sqrt{12} = \sqrt{36} = 6$
(c) $\sqrt{50} = \sqrt{25}\sqrt{2} = 5\sqrt{2}$
(d) $\sqrt{7}\sqrt{7} = \sqrt{49} = 7$

Notice from example (d) that $\sqrt{a}\sqrt{a} = a$.

Example 7(**c**) shows how Property 1 can be used to simplify certain radical expressions by removing perfect squares from under the radical.

> To simplify radicals:
>
> **1.** Factor the largest perfect square from the radicand.
> **2.** Use Property 1 to separate terms.
> **3.** Simplify.

EXAMPLE 8   Simplify $\sqrt{24}$.

SOLUTION   In 24, the largest perfect-square factor is 4.

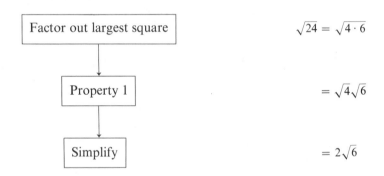

$$\sqrt{24} = \sqrt{4 \cdot 6}$$

$$= \sqrt{4}\sqrt{6}$$

$$= 2\sqrt{6}$$

EXAMPLE 9   Simplify $\sqrt{700}$.

SOLUTION   The radicand 700 has the factor 100, which is a perfect square.

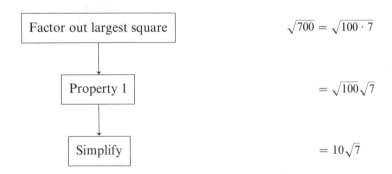

$$\sqrt{700} = \sqrt{100 \cdot 7}$$

$$= \sqrt{100}\sqrt{7}$$

$$= 10\sqrt{7}$$

EXAMPLE 10   Simplify $\sqrt{12}\sqrt{50}$.

SOLUTION   Here we simplify each radical separately then we combine.

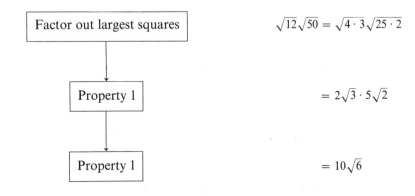

| Factor out largest squares | $\sqrt{12}\sqrt{50} = \sqrt{4\cdot 3}\sqrt{25\cdot 2}$ |

| Property 1 | $= 2\sqrt{3}\cdot 5\sqrt{2}$ |

| Property 1 | $= 10\sqrt{6}$ |

YES	NO
$\sqrt{9\cdot 16} = \sqrt{9}\sqrt{16}$	$\sqrt{9+16} = \sqrt{9}+\sqrt{16}$

Property 1 works only for products. Notice that $\sqrt{9+16} = \sqrt{25} = 5$, whereas $\sqrt{9} + \sqrt{16} = 3 + 4 = 7$. Property 1 will *not* work for sums.

There is also a property for quotients that is very similar to Property 1. Consider the following.

$$\frac{\sqrt{100}}{\sqrt{4}} = \frac{10}{2} = 5$$

$$\sqrt{\frac{100}{4}} = \sqrt{25} = 5$$

Just as $\sqrt{\dfrac{100}{4}} = \dfrac{\sqrt{100}}{\sqrt{4}}$, we have the following property.

PROPERTY 2   For positive real numbers a and b,

$$\sqrt{\frac{a}{b}} = \frac{\sqrt{a}}{\sqrt{b}}$$

EXAMPLE 11   The following examples use Property 2.

(a) $\sqrt{\dfrac{4}{25}} = \dfrac{\sqrt{4}}{\sqrt{25}} = \dfrac{2}{5}$

(b) $\sqrt{\dfrac{3}{49}} = \dfrac{\sqrt{3}}{\sqrt{49}} = \dfrac{\sqrt{3}}{7}$

(c) $\sqrt{0.09} = \sqrt{\dfrac{9}{100}} = \dfrac{\sqrt{9}}{\sqrt{100}} = \dfrac{3}{10} = 0.3$

EXAMPLE 12   Simplify $\sqrt{\dfrac{7}{18}} \cdot \dfrac{\sqrt{50}}{2}$.

SOLUTION   We simplify each radical separately, using Properties 1 and 2.

Property 2	$\sqrt{\dfrac{7}{18}} \cdot \dfrac{\sqrt{50}}{2} = \dfrac{\sqrt{7}}{\sqrt{18}} \cdot \dfrac{\sqrt{50}}{2}$
Factor out perfect squares	$= \dfrac{\sqrt{7}}{\sqrt{9 \cdot 2}} \cdot \dfrac{\sqrt{25 \cdot 2}}{2}$
Property 1	$= \dfrac{\sqrt{7}}{3\sqrt{2}} \cdot \dfrac{5\sqrt{2}}{2}$
Cancel and simplify	$= \dfrac{5\sqrt{7}}{6}$

EXAMPLE 13   The following examples involve variables (which we assume are all positive). We use Properties 1 and 2 just as we did with constants.

(a) $\sqrt{4t^4} = \sqrt{4}\sqrt{t^4} = 2t^2$

(b) $\sqrt{9w^8} = \sqrt{9}\sqrt{w^8} = 3w^4$

(c) $\sqrt{12x^7} = \sqrt{4x^6 \cdot 3x} = \sqrt{4}\sqrt{x^6}\sqrt{3x} = 2x^3\sqrt{3x}$

(d) $\sqrt{\dfrac{x^4}{y^6}} = \dfrac{\sqrt{x^4}}{\sqrt{y^6}} = \dfrac{x^2}{y^3}$

PROBLEM SET 7.2

Simplify the following radicals as much as possible using Property 1.

See
Example 7

1. $\sqrt{3}\sqrt{7}$　　　　　　　2. $\sqrt{2}\sqrt{5}$

3. $\sqrt{3}\sqrt{27}$　　　　　　4. $\sqrt{2}\sqrt{8}$

5. $\sqrt{50}\sqrt{2}$　　　　　　6. $\sqrt{32}\sqrt{2}$

7. $\sqrt{5}\sqrt{5}$　　　　　　　8. $\sqrt{7}\sqrt{7}$

9. $\sqrt{10}\sqrt{10}$　　　　　10. $\sqrt{26}\sqrt{26}$

See
Examples
8 and 9

11. $\sqrt{50}$　　　　　　　　12. $\sqrt{12}$

13. $\sqrt{18}$　　　　　　　　14. $\sqrt{48}$

15. $\sqrt{1000}$　　　　　　16. $\sqrt{80}$

17. $\sqrt{98}$　　　　　　　　18. $\sqrt{75}$

19. $\sqrt{32}$　　　　　　　　20. $\sqrt{28}$

21. $\sqrt{200}$　　　　　　　22. $\sqrt{63}$

23. $\sqrt{150}$　　　　　　　24. $\sqrt{72}$

See
Example 10

25. $\sqrt{20}\sqrt{45}$　　　　　26. $\sqrt{12}\sqrt{45}$

27. $\sqrt{75}\sqrt{32}$　　　　　28. $\sqrt{18}\sqrt{50}$

29. $\sqrt{30}\sqrt{20}$　　　　　30. $\sqrt{60}\sqrt{48}$

31. $\sqrt{200}\sqrt{28}$　　　　32. $\sqrt{40}\sqrt{60}$

Simplify the following as much as possible using Properties 1 and 2.

See
Example 11

33. $\sqrt{\dfrac{81}{100}}$　　　　　　34. $\sqrt{\dfrac{25}{9}}$

35. $\sqrt{\dfrac{25}{36}}$　　　　　　36. $\sqrt{\dfrac{1}{4}}$

37. $\sqrt{\dfrac{49}{9}}$　　　　　　38. $\sqrt{\dfrac{16}{25}}$

39. $\sqrt{\dfrac{121}{144}}$　　　　　40. $\sqrt{\dfrac{1}{100}}$

41. $\sqrt{\dfrac{7}{25}}$　　　　　　42. $\sqrt{\dfrac{11}{36}}$

43. $\sqrt{\dfrac{3}{49}}$

44. $\sqrt{\dfrac{10}{81}}$

45. $\sqrt{0.49}$

46. $\sqrt{0.04}$

47. $\sqrt{0.0081}$

48. $\sqrt{0.0001}$

See Example 12   49. $\sqrt{\dfrac{18}{8}}$

50. $\sqrt{\dfrac{12}{27}}$

51. $\sqrt{\dfrac{50}{8}}$

52. $\sqrt{\dfrac{20}{125}}$

53. $\sqrt{\dfrac{6}{15}}\sqrt{\dfrac{5}{8}}$

54. $\sqrt{\dfrac{10}{21}}\sqrt{\dfrac{3}{7}}$

55. $\sqrt{\dfrac{15}{14}}\sqrt{\dfrac{7}{2}}$

56. $\sqrt{\dfrac{20}{27}}\sqrt{\dfrac{3}{5}}$

57. $\sqrt{\dfrac{50}{49}}\sqrt{\dfrac{3}{2}}$

58. $\sqrt{\dfrac{18}{25}}\sqrt{\dfrac{2}{27}}$

Simplify the following as much as possible. (Assume that all variables are positive.)

See Example 13   59. $\sqrt{16x^{10}}$

60. $\sqrt{25r^6}$

61. $\sqrt{50p^{12}}$

62. $\sqrt{18w^2}$

63. $\sqrt{24t^5}$

64. $\sqrt{32m^3}$

65. $\sqrt{\dfrac{x^2}{9}}$

66. $\sqrt{\dfrac{k^4}{49}}$

67. $\sqrt{\dfrac{u^5}{16}}$

68. $\sqrt{\dfrac{3w^7}{25}}$

Music application   69. The overtone frequencies f_n of a vibrating string are given by

$$f_n = \frac{n}{2L}\sqrt{\frac{T}{m}}$$

where L is the length of the string, T is the tension, m is the mass density of the string, and n is a natural number $(1, 2, 3, \ldots)$. Simplify this fraction for $n = 2$, $T = 100$, $L = 20$, and $m = 0.0001$.

| 7.3 | **RADICAL EXPRESSIONS** |

In this section we look at simple **radical expressions,** such as

$$4 + \sqrt{5}, \quad 5\sqrt{7} - \sqrt{2}, \quad 2\sqrt{11} - 3, \quad \text{and} \quad 1 - \sqrt{15}$$

We want to add, subtract, and multiply these radical expressions. (We study division in Section 7.4.) The algebra of radical expressions is very similar to that of polynomials, as we saw in Chapter 3. For instance, we can treat an expression such as $3 + 2\sqrt{7}$ similarly to the way that we would treat $3 + 2x$. As with monomials, we have the idea of **like terms,** which are terms with the same radical. As examples,

$$\sqrt{5}, \ 3\sqrt{5}, \ \text{and} \ 7\sqrt{5} \ \text{are like terms}$$
$$\sqrt{6}, \ 2\sqrt{5}, \ \text{and} \ 4\sqrt{2} \ \text{are } not \text{ like terms}$$

> When we add or subtract radical expressions, we combine the like terms.

EXAMPLE 14   The following radical terms are simplified by combining like terms.

(a) $6\sqrt{2} + 7\sqrt{2} - 4\sqrt{2} = (6 + 7 - 4)\sqrt{2} = 9\sqrt{2}$
(b) $10\sqrt{5} - 8\sqrt{5} - 9\sqrt{5} = (10 - 8 - 9)\sqrt{5} = -7\sqrt{5}$

EXAMPLE 15   Add $(4 + \sqrt{3}) + (8 - 5\sqrt{3})$.

SOLUTION   We group like terms and add.

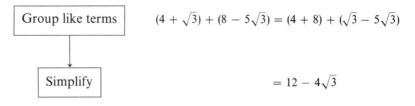

$$(4 + \sqrt{3}) + (8 - 5\sqrt{3}) = (4 + 8) + (\sqrt{3} - 5\sqrt{3})$$

$$= 12 - 4\sqrt{3}$$

Notice that we treated this exactly as we would the polynomial addition

$$(4 + x) + (8 - 5x) = 12 - 4x$$

EXAMPLE 16   Add $4\sqrt{18} + \sqrt{8}$.

SOLUTION   These do not appear to be like terms. However, by first simplifying them, they do become like terms.

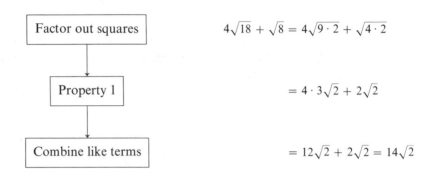

$$4\sqrt{18} + \sqrt{8} = 4\sqrt{9 \cdot 2} + \sqrt{4 \cdot 2}$$

$$= 4 \cdot 3\sqrt{2} + 2\sqrt{2}$$

$$= 12\sqrt{2} + 2\sqrt{2} = 14\sqrt{2}$$

EXAMPLE 17   Simplify $\sqrt{3a} - \sqrt{12a^3} + \sqrt{48a^5}$.

SOLUTION   We first simplify each radical, and then we combine terms.

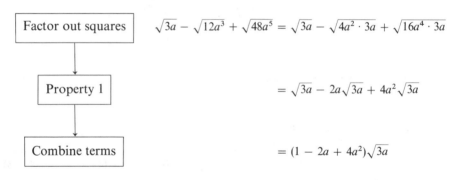

$$\sqrt{3a} - \sqrt{12a^3} + \sqrt{48a^5} = \sqrt{3a} - \sqrt{4a^2 \cdot 3a} + \sqrt{16a^4 \cdot 3a}$$

$$= \sqrt{3a} - 2a\sqrt{3a} + 4a^2\sqrt{3a}$$

$$= (1 - 2a + 4a^2)\sqrt{3a}$$

YES	NO
$\sqrt{2} + \sqrt{8} = \sqrt{2} + 2\sqrt{2} = 3\sqrt{2}$	$\sqrt{2} + \sqrt{8} = \sqrt{10}$

We must combine *like* terms to add radicals.

To multiply radical expression, we use the distributive law (or FOIL).

EXAMPLE 18   Multiply $\sqrt{2}(\sqrt{5} - \sqrt{8})$.

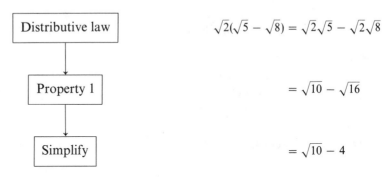

SOLUTION   We use the distributive law first; then we simplify using Property 1 that $\sqrt{a}\sqrt{b} = \sqrt{ab}$.

Distributive law	$\sqrt{2}(\sqrt{5} - \sqrt{8}) = \sqrt{2}\sqrt{5} - \sqrt{2}\sqrt{8}$
Property 1	$= \sqrt{10} - \sqrt{16}$
Simplify	$= \sqrt{10} - 4$

EXAMPLE 19   Multiply $(\sqrt{2} + 5\sqrt{3})(3\sqrt{2} - 2\sqrt{3})$.

SOLUTION   We treat this exactly as we would the polynomial multiplication $(x + 5y)(3x - 2y)$. We use the FOIL method (Firsts, Outers, Inners, and Lasts) to multiply these binomials.

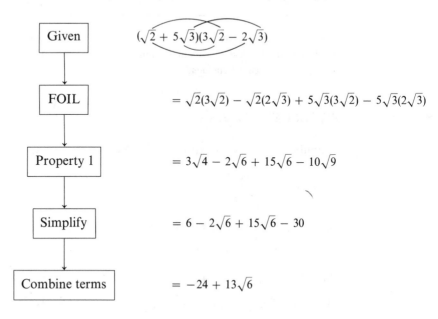

Given	$(\sqrt{2} + 5\sqrt{3})(3\sqrt{2} - 2\sqrt{3})$
FOIL	$= \sqrt{2}(3\sqrt{2}) - \sqrt{2}(2\sqrt{3}) + 5\sqrt{3}(3\sqrt{2}) - 5\sqrt{3}(2\sqrt{3})$
Property 1	$= 3\sqrt{4} - 2\sqrt{6} + 15\sqrt{6} - 10\sqrt{9}$
Simplify	$= 6 - 2\sqrt{6} + 15\sqrt{6} - 30$
Combine terms	$= -24 + 13\sqrt{6}$

EXAMPLE 20   Multiply $(4 - 3\sqrt{5})^2$.

SOLUTION   We use the special-product rule $(a - b)^2 = a^2 - 2ab + b^2$.

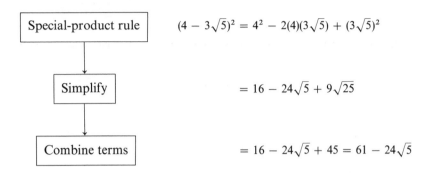

$$\boxed{\text{Special-product rule}} \qquad (4 - 3\sqrt{5})^2 = 4^2 - 2(4)(3\sqrt{5}) + (3\sqrt{5})^2$$

$$\boxed{\text{Simplify}} \qquad = 16 - 24\sqrt{5} + 9\sqrt{25}$$

$$\boxed{\text{Combine terms}} \qquad = 16 - 24\sqrt{5} + 45 = 61 - 24\sqrt{5}$$

EXAMPLE 21   This example uses the special-product rule $(a + b)(a - b) = a^2 - b^2$. Notice how this works with radicals.

	$a + b$	$a - b$	$(a + b)(a - b) = a^2 - b^2$
(a)	$3 + \sqrt{2}$	$3 - \sqrt{2}$	$9 - \sqrt{4} = 9 - 2 = ⑦$
(b)	$\sqrt{7} + 2$	$\sqrt{7} - 2$	$\sqrt{49} - 4 = 7 - 4 = ③$
(c)	$\sqrt{11} + \sqrt{5}$	$\sqrt{11} - \sqrt{5}$	$\sqrt{121} - \sqrt{25} = 11 - 5 = ⑥$
(d)	$\sqrt{2} + \sqrt{6}$	$\sqrt{2} - \sqrt{6}$	$\sqrt{4} - \sqrt{36} = 2 - 6 = ⑷$

Notice that radical expressions in this $(a + b)(a - b)$ form always have a product *without* a radical term. We use this fact in Section 7.4. Pairs of radical expressions such as $6 + \sqrt{5}$ and $6 - \sqrt{5}$ are called **conjugates.**

PROBLEM SET 7.3

Simplify the following expressions. (Assume that all variables are positive.)

See
Example 14

1. $4\sqrt{2} + 5\sqrt{2}$

2. $3\sqrt{5} + 7\sqrt{5}$

3. $10\sqrt{3} - 6\sqrt{3}$

4. $15\sqrt{7} - 8\sqrt{7}$

5. $2\sqrt{10} + 6\sqrt{10} - 3\sqrt{10}$

6. $8\sqrt{6} - 12\sqrt{6} - 7\sqrt{6}$

7. $12\sqrt{11} - 3\sqrt{11} - 16\sqrt{11}$

8. $6\sqrt{13} - 10\sqrt{13} - 7\sqrt{13}$

See
Example 15

9. $(3 + \sqrt{2}) + (6 + 4\sqrt{2})$

10. $(8 - 3\sqrt{5}) + (6 + 4\sqrt{5})$

11. $(12 - 6\sqrt{7}) - (7 - 4\sqrt{7})$

12. $(7 - 4\sqrt{2}) - (8 + 5\sqrt{2})$

13. $(\sqrt{6} - 3\sqrt{5}) + (4\sqrt{6} + 2\sqrt{5}) - (5\sqrt{6} - 8\sqrt{5})$

14. $(\sqrt{11} - \sqrt{13}) - (3\sqrt{11} + 4\sqrt{13}) - (6\sqrt{11} - 7\sqrt{13})$

15. $(\sqrt{7} - 4\sqrt{2}) - (3\sqrt{7} + 3\sqrt{2}) + (6\sqrt{7} - 2\sqrt{2})$

16. $(\sqrt{15} + \sqrt{5}) + (2\sqrt{15} - 6\sqrt{5}) - (8\sqrt{15} - 7\sqrt{5})$

See Example 16

17. $\sqrt{20} + \sqrt{45}$ 18. $\sqrt{12} + \sqrt{27}$

19. $\sqrt{200} - \sqrt{32}$ 20. $\sqrt{75} - \sqrt{48}$

21. $2\sqrt{80} - \sqrt{500}$ 22. $2\sqrt{24} + 3\sqrt{54}$

23. $3\sqrt{5} - 4\sqrt{20} + 3\sqrt{80}$ 24. $5\sqrt{3} - 3\sqrt{12} + 2\sqrt{75}$

See Example 17

25. $8\sqrt{2x} + 5\sqrt{2x}$ 26. $7\sqrt{7a} - 10\sqrt{7a}$

27. $\sqrt{5a} + \sqrt{20a^3}$ 28. $\sqrt{2t} - \sqrt{8t^3}$

29. $\sqrt{x} - 2\sqrt{x^3} + 5\sqrt{x^5}$ 30. $2\sqrt{3m} - 4\sqrt{12m^5} + 6\sqrt{27m^7}$

31. $\sqrt{50u} - 2\sqrt{32u^3} + 5\sqrt{8u^5}$ 32. $\sqrt{20k} - 3\sqrt{45k^3} + 4\sqrt{500k^9}$

See Example 18

33. $\sqrt{3}(\sqrt{5} + \sqrt{3})$ 34. $\sqrt{2}(\sqrt{7} - \sqrt{2})$

35. $\sqrt{6}(\sqrt{5} - \sqrt{2})$ 36. $\sqrt{7}(\sqrt{3} + \sqrt{5})$

37. $\sqrt{3}(\sqrt{12} + \sqrt{27})$ 38. $\sqrt{2}(\sqrt{8} - \sqrt{18})$

39. $\sqrt{5}(2\sqrt{20} - \sqrt{15})$ 40. $\sqrt{3}(\sqrt{48} - 2\sqrt{21})$

See Example 19

41. $(2 + \sqrt{3})(5 + 2\sqrt{3})$ 42. $(5 - \sqrt{2})(3 + 3\sqrt{2})$

43. $(1 - \sqrt{7})(2 + 2\sqrt{7})$ 44. $(5 + \sqrt{6})(2 + 3\sqrt{6})$

45. $(\sqrt{3} - 2\sqrt{2})(\sqrt{3} + \sqrt{2})$ 46. $(\sqrt{6} - \sqrt{2})(\sqrt{6} + 4\sqrt{2})$

47. $(2\sqrt{5} - \sqrt{3})(\sqrt{5} + \sqrt{3})$ 48. $(3\sqrt{11} - \sqrt{5})(\sqrt{11} + 4\sqrt{5})$

See Example 20

49. $(1 + \sqrt{2})^2$ 50. $(2 + \sqrt{3})^2$

51. $(5 - \sqrt{6})^2$ 52. $(2 - \sqrt{7})^2$

53. $(\sqrt{7} + \sqrt{2})^2$ 54. $(2\sqrt{5} + \sqrt{3})^2$

55. $(\sqrt{2} - \sqrt{11})^2$ 56. $(4\sqrt{3} - \sqrt{2})^2$

See Example 21

57. $(1 + \sqrt{3})(1 - \sqrt{3})$ 58. $(5 - \sqrt{2})(5 + \sqrt{2})$

59. $(\sqrt{7} + 3)(\sqrt{7} - 3)$ 60. $(2\sqrt{5} - 6)(2\sqrt{5} + 6)$

61. $(\sqrt{2} - \sqrt{3})(\sqrt{2} + \sqrt{3})$ 62. $(\sqrt{5} + \sqrt{7})(\sqrt{5} - \sqrt{7})$

63. $(2\sqrt{3} + \sqrt{11})(2\sqrt{3} - \sqrt{11})$ 64. $(\sqrt{6} - 3\sqrt{2})(\sqrt{6} + 3\sqrt{2})$

Business
application

65. To find what production level will produce the greatest profit, a company solves the equation $x^2 - 8x + 4 = 0$. They obtain a solution $4 + 2\sqrt{3}$. Check this by evaluating

$$(4 + 2\sqrt{3})^2 - 8(4 + 2\sqrt{3}) + 4$$

7.4   RATIONALIZING THE DENOMINATOR

In Section 7.3 we did not discuss the division of radical expressions. We accomplish the division of radical expressions by **simplifying** a fraction with radicals so that:

> **RULE 1**   There are no fractions under a radical sign.
> **RULE 2**   There are no radicals in a denominator.

To remove fractions from under a radical sign, use Property 2

$$\sqrt{\frac{a}{b}} = \frac{\sqrt{a}}{\sqrt{b}}$$

For example, $\sqrt{\dfrac{3}{5}} = \dfrac{\sqrt{3}}{\sqrt{5}}$. This satisfies Rule 1 by removing the fraction from under the radical sign. However, there is now a radical in the denominator (violating Rule 2). We have another procedure to cure this problem.

To simplify a fraction with a radical denominator $\sqrt{a}$:

1. Multiply the fraction by $1 = \dfrac{\sqrt{a}}{\sqrt{a}}$.

2. Simplify.

EXAMPLE 22   Simplify $\dfrac{2}{\sqrt{3}}$.

SOLUTION   Here we want to remove the $\sqrt{3}$ from the denominator. To do this we multiply the fraction by $1 = \dfrac{\sqrt{3}}{\sqrt{3}}$.

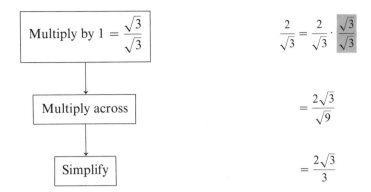

By multiplying the denominator $\sqrt{3}$ by another $\sqrt{3}$, it becomes $\sqrt{9} = 3$, which is no longer a radical. This process of removing a radical from the denominator is called **rationalizing the denominator.**

EXAMPLE 23   Simplify $\sqrt{\dfrac{5}{7}}$.

SOLUTION   We first use Property 2 to remove the fraction from under the radical sign. Then we rationalize by multiplying by $\dfrac{\sqrt{7}}{\sqrt{7}}$.

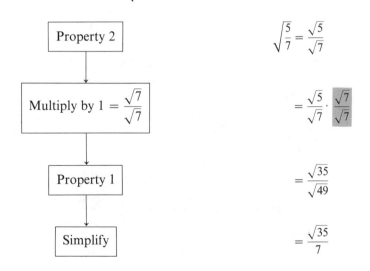

EXAMPLE 24   Simplify $\sqrt{\dfrac{y}{2x^3}}$. (Assume that x and y are positive.)

SOLUTION   We first use Property 2, and then we rationalize.

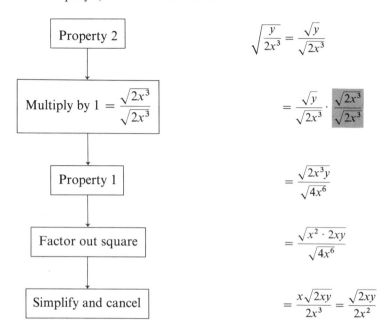

$$\sqrt{\frac{y}{2x^3}} = \frac{\sqrt{y}}{\sqrt{2x^3}}$$

$$= \frac{\sqrt{y}}{\sqrt{2x^3}} \cdot \frac{\sqrt{2x^3}}{\sqrt{2x^3}}$$

$$= \frac{\sqrt{2x^3 y}}{\sqrt{4x^6}}$$

$$= \frac{\sqrt{x^2 \cdot 2xy}}{\sqrt{4x^6}}$$

$$= \frac{x\sqrt{2xy}}{2x^3} = \frac{\sqrt{2xy}}{2x^2}$$

Sometimes, we have a radical expression in the denominator; for instance,

$$\frac{6}{4 - \sqrt{3}}$$

has a radical expression in the denominator. Recall from the preceding section (see Example 21) that multiplying conjugate pairs, such as $4 + \sqrt{3}$ and $4 - \sqrt{3}$, always gives a product without a radical.

> To simplify a fraction with a binomial radical expression in the denominator:
>
> **1.** Find the conjugate of the denominator.
>
> **2.** Multiply the fraction by $1 = \dfrac{\text{conjugate}}{\text{conjugate}}$.
>
> **3.** Simplify.

EXAMPLE 25   Simplify $\dfrac{3}{4 + \sqrt{5}}$.

SOLUTION   The conjugate of the denominator is $4 - \sqrt{5}$.

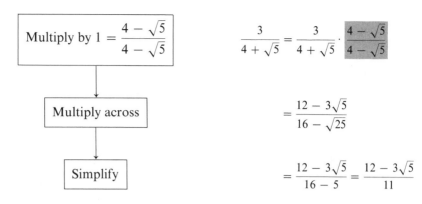

EXAMPLE 26   Simplify $\dfrac{\sqrt{2}}{\sqrt{5} - \sqrt{3}}$.

SOLUTION   The conjugate of the denominator is $\sqrt{5} + \sqrt{3}$.

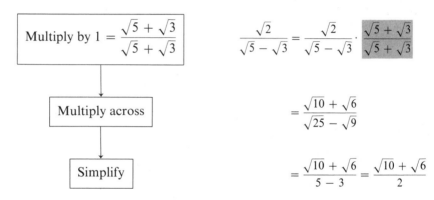

Notice that the denominator becomes a simpler number when we multiply by $\dfrac{\text{conjugate}}{\text{conjugate}}$. This process is also called **rationalizing.**

PROBLEM SET 7.4

Simplify the following as much as possible. (Assume that all variables are positive.)

See
Example 22

1. $\dfrac{5}{\sqrt{6}}$

2. $\dfrac{2}{\sqrt{7}}$

3. $\dfrac{8}{\sqrt{3}}$

4. $\dfrac{9}{\sqrt{2}}$

5. $\dfrac{16}{\sqrt{2}}$

6. $\dfrac{6}{\sqrt{3}}$

7. $\dfrac{5}{\sqrt{3}}$

8. $\dfrac{8}{\sqrt{7}}$

9. $\dfrac{\sqrt{2}}{\sqrt{11}}$

10. $\dfrac{\sqrt{13}}{\sqrt{10}}$

See Example 23   11. $\sqrt{\dfrac{1}{2}}$

12. $\sqrt{\dfrac{1}{3}}$

13. $\sqrt{\dfrac{2}{11}}$

14. $\sqrt{\dfrac{3}{7}}$

15. $\sqrt{\dfrac{7}{13}}$

16. $\sqrt{\dfrac{11}{10}}$

17. $\sqrt{\dfrac{12}{17}}$

18. $\sqrt{\dfrac{8}{11}}$

See Example 24   19. $\sqrt{\dfrac{a}{b}}$

20. $\sqrt{\dfrac{x}{y}}$

21. $\sqrt{\dfrac{x}{3y}}$

22. $\sqrt{\dfrac{2m}{3n}}$

23. $\sqrt{\dfrac{2r}{5s^3}}$

24. $\sqrt{\dfrac{8u}{3v^3}}$

25. $\sqrt{\dfrac{10a^3}{7b^5}}$

26. $\sqrt{\dfrac{14}{5t^9}}$

See Example 25   27. $\dfrac{4}{3+\sqrt{2}}$

28. $\dfrac{5}{6-\sqrt{3}}$

29. $\dfrac{1}{2-\sqrt{3}}$

30. $\dfrac{10}{5+\sqrt{7}}$

31. $\dfrac{2}{\sqrt{2}-7}$

32. $\dfrac{4}{\sqrt{5}+7}$

33. $\dfrac{\sqrt{2}}{\sqrt{11}+3}$

34. $\dfrac{\sqrt{3}}{\sqrt{10}-5}$

See Example 26   35. $\dfrac{1}{\sqrt{3}-\sqrt{2}}$

36. $\dfrac{3}{\sqrt{5}+\sqrt{2}}$

37. $\dfrac{4}{\sqrt{7} + \sqrt{2}}$

38. $\dfrac{4}{\sqrt{11} - \sqrt{3}}$

39. $\dfrac{3}{\sqrt{5} - \sqrt{2}}$

40. $\dfrac{7}{\sqrt{6} + \sqrt{2}}$

41. $\dfrac{\sqrt{10}}{\sqrt{15} + \sqrt{5}}$

42. $\dfrac{\sqrt{11}}{\sqrt{22} - \sqrt{2}}$

Chemistry application **43.** The increase in temperature, ΔT, due to a catalyst of concentration u is given by

$$\Delta T = \frac{10\sqrt{u}}{\sqrt{u} + 1}$$

Simplify this fraction.

Business application **44.** A marketing firm discovers that the time t (in months) that it takes for their product to be known by a fraction f of a community is given by

$$t = \frac{1}{1 - \sqrt{f}}$$

Simplify this fraction.

Health application **45.** The pulse rate P in beats per minute is approximated by

$$P = \frac{600}{\sqrt{h}}$$

where h is a person's height in inches. Simplify this fraction.

Psychology application **46.** The rate at which a new worker learns certain tasks is given by

$$\text{rate} = \frac{30}{\sqrt{t}}$$

where t is the time in hours. Simplify this fraction.

7.5 EQUATIONS INVOLVING RADICALS

Now we look at two types of equations that involve radicals:

1. Equations with variables under a radical sign.

2. Equations with radicals in the solution.

The first type of equation has a *variable under a radical sign*. For instance, $\sqrt{x + 5} = 4$ is such an equation. To solve these equations, we use the **squaring property of equality.**

> To solve equations with variables under a radical:
>
> **1.** Square both sides of the equation.
> **2.** Solve the resulting equation.
> **3.** Check the results to see that they are really solutions (some may not be).

When working these problems, recall the following property.

> ***PROPERTY 3***   For any positive real number N,
>
> $$(\sqrt{N})^2 = N$$

EXAMPLE 27   Solve $\sqrt{2x + 4} = 6$.

SOLUTION   We square both sides to eliminate the radical. We use Property 3, $(\sqrt{N})^2 = N$.

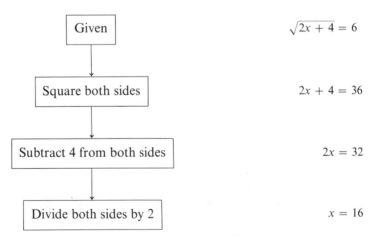

Given	$\sqrt{2x + 4} = 6$
Square both sides	$2x + 4 = 36$
Subtract 4 from both sides	$2x = 32$
Divide both sides by 2	$x = 16$

We now check this by substituting $x = 16$ into the original equation.

Substitute $x = 16$	$\sqrt{2(16) + 4} = \sqrt{36} = 6$ *Checks.*

EXAMPLE 28   Solve $\sqrt{2a + 1} = \sqrt{3a - 5}$.

SOLUTION   Again we square both sides to eliminate the radical.

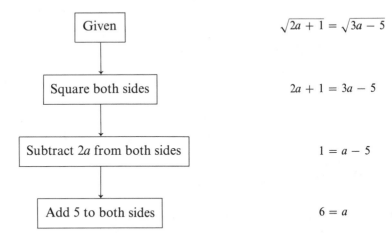

| Given | $\sqrt{2a + 1} = \sqrt{3a - 5}$ |

Square both sides $2a + 1 = 3a - 5$

Subtract $2a$ from both sides $1 = a - 5$

Add 5 to both sides $6 = a$

Let us check $a = 6$ in the original equation.

$$Left\ side\ = \sqrt{2(6) + 1} = \sqrt{13}$$
$$Right\ side = \sqrt{3(6) - 5} = \sqrt{13} \quad Checks.$$

EXAMPLE 29   Solve $\sqrt{x - 3} = -4$.

SOLUTION   We square both sides.

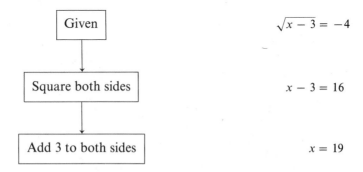

Given $\sqrt{x - 3} = -4$

Square both sides $x - 3 = 16$

Add 3 to both sides $x = 19$

Let us check $x = 19$ in the original equation.

$$Left\ side = \sqrt{19 - 3} = \sqrt{16} = 4$$

But the right side of the original equation is -4. This does *not* check! Therefore, there is *no* solution. This example shows how important it is to check.

The other type of equation that we consider here involves a *radical within the solution*. To solve these, we use the following property.

PROPERTY 4 If $x^2 = a$, then

$$x = \sqrt{a} \quad \text{or} \quad x = -\sqrt{a} \quad \text{(for } a \geq 0\text{)}$$

In words, we can *take the square roots of both sides of the equation* (if both sides of the equation are positive).

EXAMPLE 30 Solve $x^2 = 49$.

SOLUTION We simply take the square roots of both sides of the equation and get $x = 7$ or $x = -7$. The solution set is $\{7, -7\}$.

EXAMPLE 31 Solve $(a + 4)^2 = 25$.

SOLUTION We take the square roots of both sides of the equation and solve the resulting equation.

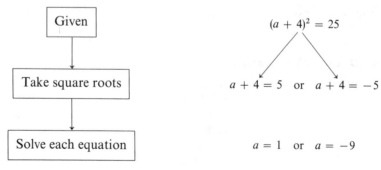

Let us check this solution set $\{1, -9\}$.

| Substitute $a = 1$ | | $(1 + 4)^2 = 5^2 = 25$ | *Checks.* |

| Substitute $a = -9$ | | $(-9 + 4)^2 = (-5)^2 = 25$ | *Checks.* |

EXAMPLE 32   Solve $(3k - 2)^2 = 15$.

SOLUTION   Again, we take the square roots of both sides. This time, however, we will get a radical in the solution.

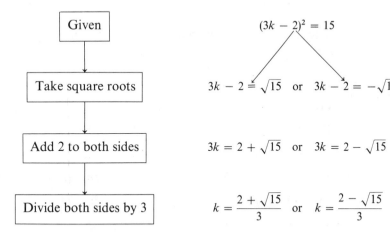

The solution set is $\left\{ \dfrac{2 + \sqrt{15}}{3}, \dfrac{2 - \sqrt{15}}{3} \right\}$.

YES	NO
$x^2 = 21$	$x^2 = 21$
$x = -\sqrt{21}$ or $x = \sqrt{21}$	$x = \sqrt{21}$

Do not forget the negative square root.

PROBLEM SET 7.5

Solve the following equations. (Be sure to check all solutions in the original equations—see Example 29.)

See
Example 27

1. $\sqrt{x + 1} = 5$ 2. $\sqrt{a - 2} = 6$

3. $\sqrt{4 - r} = 10$ 4. $\sqrt{20 - t} = 2$

5. $\sqrt{2x + 1} = 3$ 6. $\sqrt{3k + 1} = 7$

7. $\sqrt{5m - 1} = -8$ 8. $\sqrt{4u - 3} = 5$

9. $\sqrt{1 - 3m} = 4$ 10. $\sqrt{1 - 5v} = -6$

See
Example 28
11. $\sqrt{x + 1} = \sqrt{2x - 1}$ **12.** $\sqrt{a + 4} = \sqrt{2a - 2}$

13. $\sqrt{2x + 6} = \sqrt{5x}$ **14.** $\sqrt{x + 5} = \sqrt{3x - 15}$

15. $\sqrt{4x + 2} = -\sqrt{7x - 13}$ **16.** $\sqrt{5t + 2} = \sqrt{2t - 7}$

17. $\sqrt{1 + t} = \sqrt{6t - 9}$ **18.** $\sqrt{k + 10} = \sqrt{2 - k}$

Solve the following equations. (*Be careful;* some may have no real solution—see Example 3.)

See
Example 30
19. $x^2 = 1$ **20.** $k^2 = 121$

21. $m^2 = 4$ **22.** $u^2 = 64$

23. $t^2 = -16$ **24.** $y^2 = 25$

25. $w^2 = 100$ **26.** $p^2 = -49$

See
Example 31
27. $(x - 5)^2 = 25$ **28.** $(t + 4)^2 = 36$

29. $(10 - u)^2 = 1$ **30.** $(7 - m)^2 = 49$

31. $(2x + 3)^2 = 121$ **32.** $(3k - 2)^2 = -16$

33. $(2m - 3)^2 = 49$ **34.** $(2 - 4x)^2 = 100$

See
Example 32
35. $(x + 5)^2 = 6$ **36.** $(t - 3)^2 = 10$

37. $(y - 4)^2 = 3$ **38.** $(k + 7)^2 = 2$

39. $(2x + 1)^2 = 11$ **40.** $(5r - 6)^2 = 13$

41. $(7k - 10)^2 = -8$ **42.** $(3m + 5)^2 = 7$

*Engineering
application*
43. The horizontal distance d (in miles) that an antenna can project a signal is given by

$$d = 1.22 \sqrt{h}$$

where h is the height (in feet) of the antenna. Find the height h needed to project a signal $d = 20$ miles.

7.6   THE QUADRATIC FORMULA

In Chapter 4 we studied the **quadratic equation,** which has the **standard form**

$$ax^2 + bx + c = 0 \quad (a \neq 0)$$

We solved these equations by factoring the polynomial on the left, setting each factor equal to zero, and solving.

But what if the polynomial cannot be factored? In those cases we have a remarkable formula for solving any quadratic equation. We now show how this formula is derived. (The trick is to make the left side of the equation into a perfect square.)

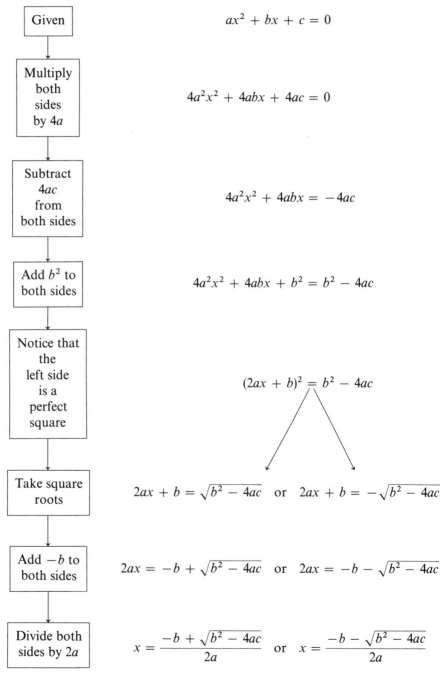

| Given | $ax^2 + bx + c = 0$ |

Multiply both sides by $4a$
$$4a^2x^2 + 4abx + 4ac = 0$$

Subtract $4ac$ from both sides
$$4a^2x^2 + 4abx = -4ac$$

Add b^2 to both sides
$$4a^2x^2 + 4abx + b^2 = b^2 - 4ac$$

Notice that the left side is a perfect square
$$(2ax + b)^2 = b^2 - 4ac$$

Take square roots
$$2ax + b = \sqrt{b^2 - 4ac} \quad \text{or} \quad 2ax + b = -\sqrt{b^2 - 4ac}$$

Add $-b$ to both sides
$$2ax = -b + \sqrt{b^2 - 4ac} \quad \text{or} \quad 2ax = -b - \sqrt{b^2 - 4ac}$$

Divide both sides by $2a$
$$x = \frac{-b + \sqrt{b^2 - 4ac}}{2a} \quad \text{or} \quad x = \frac{-b - \sqrt{b^2 - 4ac}}{2a}$$

There are generally *two* solutions, and we usually write them in a compact form known as the **quadratic formula.**

THEOREM 1   (*Quadratic Formula*)   The solutions to the equation $ax^2 + bx + c = 0$ $(a \neq 0)$ are

$$x = \frac{-b \pm \sqrt{b^2 - 4ac}}{2a}$$

To use the quadratic formula:

1. Put the equation into standard form $ax^2 + bx + c = 0$.
2. Identify the terms a, b, and c.
3. Substitute into the quadratic formula and simplify. (Be careful with negative numbers.)

EXAMPLE 33   Solve $x^2 - 5x + 6 = 0$ using the quadratic formula.

SOLUTION   The equation is already in standard form.

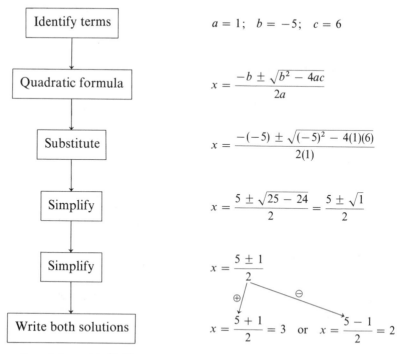

| Identify terms | $a = 1; \quad b = -5; \quad c = 6$ |

| Quadratic formula | $x = \dfrac{-b \pm \sqrt{b^2 - 4ac}}{2a}$ |

| Substitute | $x = \dfrac{-(-5) \pm \sqrt{(-5)^2 - 4(1)(6)}}{2(1)}$ |

| Simplify | $x = \dfrac{5 \pm \sqrt{25 - 24}}{2} = \dfrac{5 \pm \sqrt{1}}{2}$ |

| Simplify | $x = \dfrac{5 \pm 1}{2}$ |

| Write both solutions | $x = \dfrac{5 + 1}{2} = 3 \quad$ or $\quad x = \dfrac{5 - 1}{2} = 2$ |

Thus, the solution set is $\{2, 3\}$.

EXAMPLE 34   Solve $2x^2 - x = 15$.

SOLUTION   We first put this into standard form, $2x^2 - x - 15 = 0$.

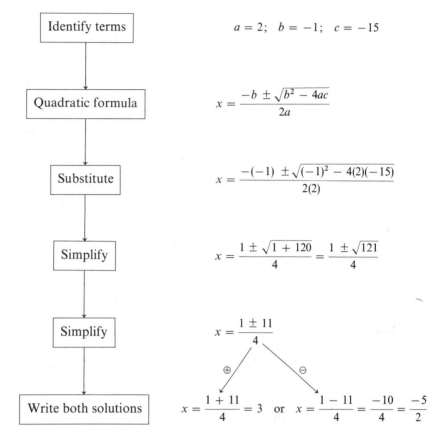

Identify terms $a = 2;\quad b = -1;\quad c = -15$

Quadratic formula $x = \dfrac{-b \pm \sqrt{b^2 - 4ac}}{2a}$

Substitute $x = \dfrac{-(-1) \pm \sqrt{(-1)^2 - 4(2)(-15)}}{2(2)}$

Simplify $x = \dfrac{1 \pm \sqrt{1 + 120}}{4} = \dfrac{1 \pm \sqrt{121}}{4}$

Simplify $x = \dfrac{1 \pm 11}{4}$

Write both solutions $x = \dfrac{1 + 11}{4} = 3 \quad \text{or} \quad x = \dfrac{1 - 11}{4} = \dfrac{-10}{4} = \dfrac{-5}{2}$

The solution set is $\left\{ 3, \dfrac{-5}{2} \right\}$.

EXAMPLE 35   Solve $\dfrac{2}{9}x^2 + x - \dfrac{2}{3} = 0$.

SOLUTION   This problem would be very ugly to do with the fractions, so we clear the fractions as we did in Chapter 5: multiply both sides of the equation by the LCD, 9. Then we use the quadratic formula. (Recall that $9 \cdot 0 = 0$.)

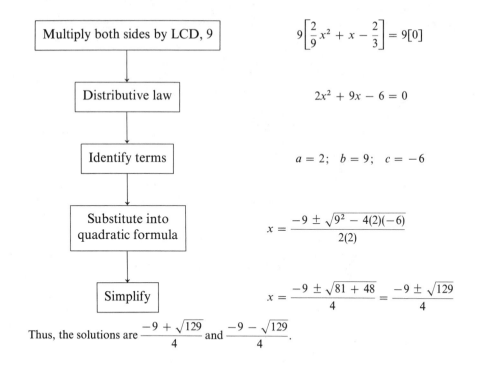

| Multiply both sides by LCD, 9 | $9\left[\dfrac{2}{9}x^2 + x - \dfrac{2}{3}\right] = 9[0]$ |

| Distributive law | $2x^2 + 9x - 6 = 0$ |

| Identify terms | $a = 2; \quad b = 9; \quad c = -6$ |

| Substitute into quadratic formula | $x = \dfrac{-9 \pm \sqrt{9^2 - 4(2)(-6)}}{2(2)}$ |

| Simplify | $x = \dfrac{-9 \pm \sqrt{81 + 48}}{4} = \dfrac{-9 \pm \sqrt{129}}{4}$ |

Thus, the solutions are $\dfrac{-9 + \sqrt{129}}{4}$ and $\dfrac{-9 - \sqrt{129}}{4}$.

EXAMPLE 36   Solve $(x - 1)(x - 4) = 3x$.

SOLUTION   We first multiply the two binomials and then put the equation into standard form.

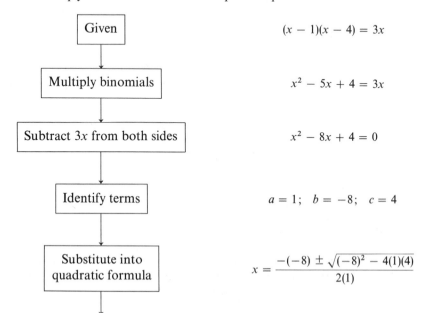

| Given | $(x - 1)(x - 4) = 3x$ |

| Multiply binomials | $x^2 - 5x + 4 = 3x$ |

| Subtract $3x$ from both sides | $x^2 - 8x + 4 = 0$ |

| Identify terms | $a = 1; \quad b = -8; \quad c = 4$ |

| Substitute into quadratic formula | $x = \dfrac{-(-8) \pm \sqrt{(-8)^2 - 4(1)(4)}}{2(1)}$ |

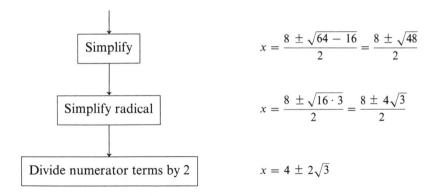

$$x = \frac{8 \pm \sqrt{64 - 16}}{2} = \frac{8 \pm \sqrt{48}}{2}$$

$$x = \frac{8 \pm \sqrt{16 \cdot 3}}{2} = \frac{8 \pm 4\sqrt{3}}{2}$$

$$x = 4 \pm 2\sqrt{3}$$

Thus, the solutions are $4 + 2\sqrt{3}$ and $4 - 2\sqrt{3}$. Equivalently, we can say that the solution set is $\{4 + 2\sqrt{3}, 4 - 2\sqrt{3}\}$.

EXAMPLE 37   Write the solutions to Example 35 as decimals.

SOLUTION   Recall that the solutions are $\dfrac{-9 + \sqrt{129}}{4}$ and $\dfrac{-9 - \sqrt{129}}{4}$. We compute these using a hand calculator. We compute the "+" solution first and then the "−" solution.

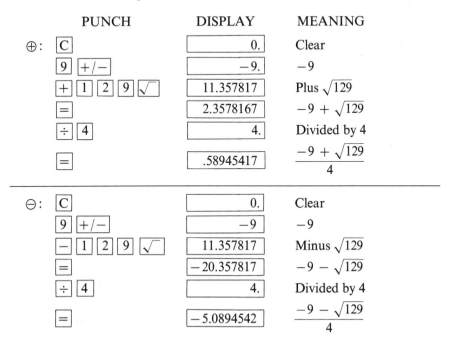

Thus, the solutions are approximately 0.589 and -5.089.

PROBLEM SET 7.6

Use the quadratic formula to solve the following equations.

See
Example 33

1. $x^2 + 7x + 10 = 0$ **2.** $m^2 + 7m - 8 = 0$

3. $y^2 - 3y - 18 = 0$ **4.** $a^2 + 14a + 40 = 0$

5. $t^2 + 5t + 4 = 0$ **6.** $r^2 - 11r + 30 = 0$

7. $x^2 + x - 6 = 0$ **8.** $m^2 + 9m + 14 = 0$

9. $a^2 - 11a + 10 = 0$ **10.** $u^2 + 7u - 18 = 0$

See
Example 34

11. $6x^2 + x = 1$ **12.** $6u^2 + 15 = 19u$

13. $8a^2 + 22a = -5$ **14.** $10m^2 + 13m = 3$

15. $2x^2 + 5x = 3$ **16.** $3k^2 = 7k + 1$

17. $x^2 + 9x = 4$ **18.** $2a^2 + 1 = 5a$

19. $5x^2 = 3x + 1$ **20.** $4t^2 = t + 7$

See
Example 35

21. $\dfrac{x^2}{5} - \dfrac{x}{15} - \dfrac{1}{3} = 0$ **22.** $\dfrac{a^2}{10} + \dfrac{a}{4} - \dfrac{1}{5} = 0$

23. $\dfrac{x^2}{2} + \dfrac{x}{10} - \dfrac{1}{5} = 0$ **24.** $\dfrac{t^2}{3} - \dfrac{t}{3} - \dfrac{1}{4} = 0$

25. $\dfrac{m^2}{8} + \dfrac{1}{4} = m$ **26.** $\dfrac{k^2}{5} + 1 = k$

27. $\dfrac{u^2}{6} = \dfrac{1}{4} - \dfrac{u}{3}$ **28.** $\dfrac{y^2}{5} = \dfrac{1}{3} - \dfrac{y}{15}$

29. $\dfrac{x^2}{12} = \dfrac{1}{3} + \dfrac{x}{4}$ **30.** $\dfrac{z^2}{5} = \dfrac{z}{2} - \dfrac{1}{10}$

See
Example 36

31. $(x + 1)(x - 1) = 4$ **32.** $(k - 2)(k - 3) = 7$

33. $(r + 2)(r - 3) = 2r$ **34.** $(m + 1)(m + 1) = 10m$

35. $(n + 5)(n - 1) = 9n$ **36.** $(k + 2)(k + 4) = 13k$

37. $(2x - 1)(x + 5) = 7x$ **38.** $(a - 4)(2a + 1) = -8$

Write the solutions to the following problems as three-place decimals. (Use a hand calculator or a table of square roots.)

See
Example 37

39. Problem 17   **40.** Problem 18

41. Problem 21   **42.** Problem 26

43. Problem 27

44. Problem 30

45. Problem 33

46. Problem 36

Business application **47.** The supply S and demand D for a certain box of candy are given by

$$S = 30{,}000p - 20{,}000 \quad \text{and} \quad D = \frac{90{,}000}{p}$$

where p is the price. Find the *equilibrium price* where $S = D$ by solving

$$30{,}000p - 20{,}000 = \frac{90{,}000}{p}$$

(*Hint:* Put the equation into standard form; then divide both sides by 10,000 before using the quadratic formula. Also, take only the larger of the solutions.)

Health application **48.** The amount of sugar S in the blood after a certain ingestion of food is given by $S = 3.8 + 0.2t - 0.1t^2$, where t is the time in hours. Find the time at which the sugar level is 0.2 by solving the equation

$$3.8 + 0.2t - 0.1t^2 = 0.2$$

(*Hint:* Put the equation into standard form; then multiply both sides by -10 before using the quadratic formula. Also, take only the larger solution.)

Art application **49.** The *golden rectangle* shown has dimensions 1 by x. If a 1 by 1 square is cut out, the remaining rectangle has dimensions $(x - 1)$ by 1. It is related to the original rectangle by the equation

$$\frac{\text{length}}{\text{width}} = \frac{x}{1} = \frac{1}{x - 1}$$

Solve this for x. (Take only the positive solution.)

CHAPTER 7 SUMMARY

Important Words and Phrases

conjugates (7.3)

cube root (7.1)

fourth root (7.1)

index (7.1)

irrational number (7.1)

like terms (7.3)

Important Properties

For real numbers $a \geq 0$ and $b \geq 0$,

$$\sqrt{ab} = \sqrt{a}\sqrt{b}$$

$$\sqrt{a}\sqrt{a} = (\sqrt{a})^2 = a$$

$$\sqrt{\frac{a}{b}} = \frac{\sqrt{a}}{\sqrt{b}} \quad (b \neq 0)$$

A radical expression is simplified if:

1. No perfect-square factors are under the radical sign.
2. No fractions are under the radical sign.
3. No radicals are in the denominator.

We can square both sides of an equation to find its solution, but we must check all the results in the original equation.

If $x^2 = a$, then $x = \sqrt{a}$ or $x = -\sqrt{a}$ (for $a \geq 0$).

The solutions to $ax^2 + bx + c = 0$ are given by

$$x = \frac{-b \pm \sqrt{b^2 - 4ac}}{2a}$$

Important Procedures

To simplify radicals:

1. Factor the largest perfect square out of the radicand.
2. Use the property that $\sqrt{ab} = \sqrt{a}\sqrt{b}$ and simplify.

To add radical expressions, combine like terms.

To multiply radical expressions, use the distributive law or FOIL.

To remove a fraction from under a radical sign, use $\sqrt{\dfrac{a}{b}} = \dfrac{\sqrt{a}}{\sqrt{b}}$.

To simplify a fraction with a radical denominator, $\sqrt{a}$, multiply by $1 = \dfrac{\sqrt{a}}{\sqrt{a}}$.

To simplify a fraction with a binomial radical expression in the denominator:

1. Find the conjugate of the denominator.

2. Multiply by $1 = \dfrac{\text{conjugate}}{\text{conjugate}}$.

3. Simplify.

CHAPTER 7 REVIEW EXERCISES

Find the indicated numbers. (If no such real number exists, state this.)

1. The square roots of 81.   **2.** The square roots of -16.

3. $\sqrt{36}$ **4.** $-\sqrt{49}$

5. $\sqrt[3]{27}$ **6.** $\sqrt[4]{16}$

State whether the following numbers are rational or irrational.

7. $\sqrt{25}$ **8.** $\sqrt{26}$

Simplify the following as much as possible. (Assume that all variables are positive.)

9. $\sqrt{6}\sqrt{6}$ **10.** $\sqrt{125}$

11. $\sqrt{18}\sqrt{32}$ **12.** $\sqrt{\dfrac{11}{16}}$

13. $\sqrt{0.16}$ **14.** $\sqrt{\dfrac{40}{32}}$

15. $\sqrt{\dfrac{x^3}{25}}$

16. $3\sqrt{5} - 6\sqrt{5} - 7\sqrt{5}$

17. $(4 + 3\sqrt{2}) - (5 - 7\sqrt{2})$

18. $\sqrt{12} + \sqrt{75}$

19. $\sqrt{3x} - \sqrt{27x^3}$

20. $\sqrt{5}(\sqrt{7} - \sqrt{2})$

21. $(\sqrt{5} - \sqrt{2})(\sqrt{5} + 3\sqrt{2})$

22. $(\sqrt{3} - 2)^2$

23. $(5 - \sqrt{2})(5 + \sqrt{2})$

24. $\dfrac{\sqrt{5}}{\sqrt{7}}$

25. $\sqrt{\dfrac{7}{11}}$

26. $\sqrt{\dfrac{2x}{3y^3}}$

27. $\dfrac{3}{5 - \sqrt{2}}$

28. $\dfrac{\sqrt{2}}{\sqrt{7} + \sqrt{3}}$

Solve the following equations.

29. $\sqrt{2x - 5} = 7$

30. $\sqrt{k + 1} = \sqrt{3k - 1}$

31. $x^2 = 81$

32. $(4a - 1)^2 = 4$

33. $(2x - 3)^2 = 7$

34. $2x^2 + 7x + 6 = 0$

35. $4x^2 = 5x + 3$

36. $\dfrac{x^2}{12} - \dfrac{x}{2} + \dfrac{1}{3} = 0$

37. $(u - 3)(u + 2) = 3$

38. Express the solutions to problem 36 as three-place decimals.

8

ADDITIONAL GRAPHING TOPICS

8.1 SLOPE

It is often important to know how steep a line is (or how rapidly it is rising or falling to the right). To help us measure this steepness, we compute the **slope** (denoted by **m**). It is defined as follows.

DEFINITION 1 The slope of the line through points (x_1, y_1) and (x_2, y_2) is given by

$$m = \frac{\text{rise}}{\text{run}} = \frac{\text{change in } y}{\text{change in } x} = \frac{y_2 - y_1}{x_2 - x_1}$$

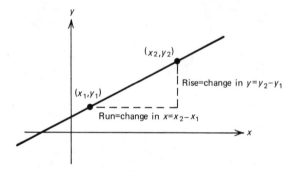

Figure 8.1

Figure 8.1 illustrates this idea. We are comparing (with a ratio) the change in y (the vertical rise) to the change in x (the horizontal run). To find these changes, we subtract the y-coordinates and subtract the x-coordinates. (When the co-ordinates are negative, special caution must be given to the subtraction.)

EXAMPLE 1 Find the slope of the line through (1, 3) and (6, 7), as shown in Figure 8.2(a).

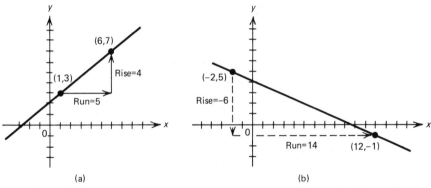

Figure 8.2

SOLUTION We first compute the changes in y and x.

$$\text{Rise} = \text{change in } y = 7 - 3 = 4$$

$$\text{Run} = \text{change in } x = 6 - 1 = 5$$

Thus,

Definition of slope	$m = \dfrac{\text{change in } y}{\text{change in } x} = \dfrac{7 - 3}{6 - 1} = \dfrac{4}{5}$

The slope is $\dfrac{4}{5}$.

EXAMPLE 2 Find the slope of the line through $(-2, 5)$ and $(12, -1)$ as shown in Figure 8.2(b).

SOLUTION The coordinates involve negative numbers, so we must subtract them carefully.

Definition of slope	$m = \dfrac{\text{change in } y}{\text{change in } x} = \dfrac{y_2 - y_1}{x_2 - x_1}$
Substitute and simplify	$m = \dfrac{-1 - 5}{12 - (-2)} = \dfrac{-6}{14} = \dfrac{-3}{7}$

EXAMPLE 3 Find the slope of the line through $(-5, -4)$ and $(6, -1)$ as shown in Figure 8.3.

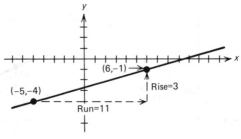

Figure 8.3

SOLUTION   Again, we work very carefully with the negatives when we subtract.

| Definition of slope | $m = \dfrac{\text{change in } y}{\text{change in } x}$ |

| Substitute and simplify | $m = \dfrac{-1 - (-4)}{6 - (-5)} = \dfrac{3}{11}$ |

YES	NO
For points $(-5, -4)$ and $(6, -1)$,	~~For points $(-5, -4)$ and $(6, -1)$,~~
$m = \dfrac{-1 - (-4)}{6 - (-5)} = \dfrac{3}{11}$	~~$m = \dfrac{-1 - 4}{6 - 5} = \dfrac{-5}{1}$~~

Be sure to *subtract* the coordinates. Often, this may involve changing the signs and adding.

EXAMPLE 4   Find the slopes of the lines, l_1 and l_2, shown in Figure 8.4, using the indicated points.

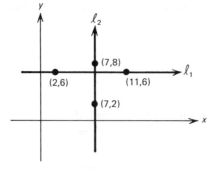

Figure 8.4

SOLUTION   We first compute the slope of l_1 and then l_2.

$$m_1 = \frac{\text{change in } y}{\text{change in } x}$$

$$m_1 = \frac{6 - 6}{11 - 2} = \frac{0}{9} = 0$$

The slope of l_1 is 0. In fact, all horizontal lines have slope 0.

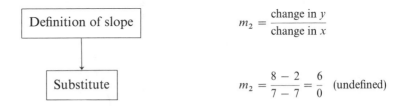

$$m_2 = \frac{\text{change in } y}{\text{change in } x}$$

$$m_2 = \frac{8 - 2}{7 - 7} = \frac{6}{0} \quad \text{(undefined)}$$

There is *no* slope for l_2, since $\frac{6}{0}$ is undefined. In fact, all vertical lines have an undefined slope.

In Examples 1 and 3 the lines had positive slopes, and the lines went up to the right. In Example 2 the line had a negative slope, and it went down to the right. Table 8.1 summarizes the facts that we have seen about slopes.

Table 8.1

LINE	SLOPE
Rising	Positive
Falling	Negative
Horizontal	Zero
Vertical	Undefined

We can use the slope to help us graph lines.

EXAMPLE 5   Graph the line through (1, 2) with slope 3.

SOLUTION   A slope of 3 means that the y-coordinate changes three times faster than the x-coordinate. For instance:

$$\text{If } x \text{ changes } 1, y \text{ changes } 3$$

$$\text{If } x \text{ changes } 2, y \text{ changes } 6$$

$$\text{If } x \text{ changes } 3, y \text{ changes } 9$$

and so on

To graph this line, we start at the point $(1, 2)$. Then we move away from $(1, 2)$ using the changes computed above.

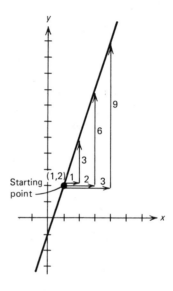

EXAMPLE 6   Graph the line through $(-1, 6)$ with slope $\dfrac{-1}{2}$.

SOLUTION   Here, a slope of $\dfrac{-1}{2}$ means that y goes *down* $\dfrac{1}{2}$ every time x goes *up* 1.

$$\text{If } x \text{ changes } 1, y \text{ changes } \frac{-1}{2}$$

$$\text{If } x \text{ changes } 2, y \text{ changes } \frac{-2}{2} (= -1)$$

$$\text{If } x \text{ changes } 3, y \text{ changes } \frac{-3}{2}$$

and so on

Now we start at $(-1, 6)$ and make these changes.

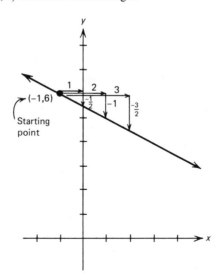

PROBLEM SET 8.1

Compute the slopes of the lines through the following points.

See
Examples
1 to 4

1. $(1, 5)$ and $(7, 12)$ **2.** $(8, 3)$ and $(10, 8)$

3. $(1, 7)$ and $(4, 6)$ **4.** $(2, 9)$ and $(7, 4)$

5. $(-1, 4)$ and $(3, 6)$ **6.** $(5, -2)$ and $(7, -5)$

7. $(-3, 6)$ and $(-8, -1)$ **8.** $(4, -1)$ and $(5, 7)$

9. $(-2, -1)$ and $(8, -6)$ **10.** $(5, -2)$ and $(-8, -5)$

11. $(7, -2)$ and $(7, 4)$ **12.** $(8, -4)$ and $(-3, -4)$

13. $(-2, 5)$ and $(9, 5)$ **14.** $(-3, 5)$ and $(-3, -3)$

Graph the lines with the given points and slopes.

See
Examples
5 and 6

15. $(2, 3)$; $m = 2$ **16.** $(5, 1)$; $m = 3$

17. $(-1, 2)$; $m = 1$ **18.** $(2, -3)$; $m = \dfrac{1}{2}$

19. $(3, 8)$; $m = -2$ **20.** $(5, -2)$; $m = -3$

21. $(-2, -3)$; $m = -1$ **22.** $(-4, -5)$; $m = \dfrac{-3}{2}$

Business application   **23.** The *marginal profit MP* is given by

$$MP = \frac{P_2 - P_1}{Q_2 - Q_1}$$

where P is the profit and Q is the quantity produced. Compute the marginal profit for the following (Q, P) pairs.

(a) (150; \$35,000) and (175; \$40,000)
(b) (3000; \$900,000) and (3200; \$1,000,000)
(c) (5000; \$200,000) and (5100; \$180,000)

Science application   **24.** The average velocity v of a moving object is given by

$$v = \frac{s_2 - s_1}{t_2 - t_1}$$

where s is the distance that an object travels in time t. Compute the average velocity for the following (t, s) pairs.

(a) (20 seconds, 300 meters) and (25 seconds, 400 meters)
(b) (3 days, 1800 miles) and (5 days, 2700 miles)
(c) (6 hours, 320 miles) and (7.5 hours, 395 miles)

8.2   THE EQUATION OF A STRAIGHT LINE

In Section 6.3 we saw that a straight line can be written as a linear equation, such as $2x + 5y = 10$. In Section 6.3 we were given the equation, and we wanted the graph. In this section we are given some information about the line, and we want the linear equation for the line.

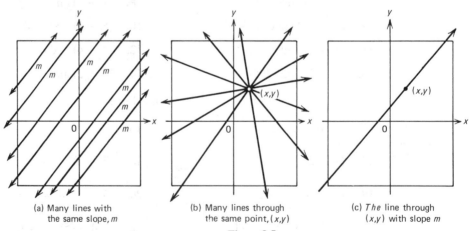

(a) Many lines with the same slope, m

(b) Many lines through the same point, (x,y)

(c) *The* line through (x,y) with slope m

Figure 8.5

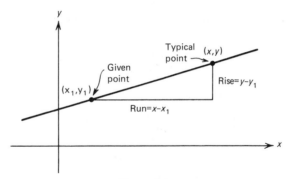

Figure 8.6

The most useful information is a point on the line and the slope. Figure 8.5(a) and (b) show what happens if we have only the slope or only a point: we need *both* the slope and a point to find the line [as shown in Figure 8.5(c)].

Figure 8.6 illustrates the situation that we have: We are given a point (x_1, y_1) and a slope m, and we want to find the equation or relation for the typical point (x, y). We start with the definition of the slope.

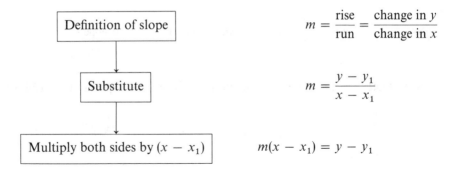

This becomes a very valuable theorem.

THEOREM 1   The equation of the line through (x_1, y_1) with slope m is given by

$$y - y_1 = m(x - x_1)$$

This is called the **point-slope form,** since a point and the slope are given.

EXAMPLE 7   Find the equation of the line with slope 3 and through (1, 7).

SOLUTION   We substitute the slope and point into the point-slope form.

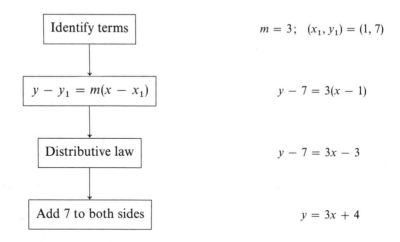

Identify terms	$m = 3;\ (x_1, y_1) = (1, 7)$
$y - y_1 = m(x - x_1)$	$y - 7 = 3(x - 1)$
Distributive law	$y - 7 = 3x - 3$
Add 7 to both sides	$y = 3x + 4$

This is the equation of the line: $y = 3x + 4$.

EXAMPLE 8   Find the equation of the line through $(2, -3)$ with slope -5.

SOLUTION   Again we substitute into the point-slope form. (We must be careful subtracting.)

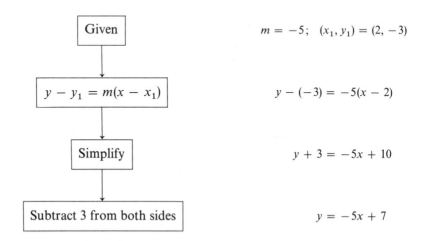

Given	$m = -5;\ (x_1, y_1) = (2, -3)$
$y - y_1 = m(x - x_1)$	$y - (-3) = -5(x - 2)$
Simplify	$y + 3 = -5x + 10$
Subtract 3 from both sides	$y = -5x + 7$

This is the equation of the line: $y = -5x + 7$.

EXAMPLE 9   Find the equation of the line through points $(-1, 2)$ and $(1, 8)$.

SOLUTION   We need a point and the slope. We have a point (two, in fact), but we do not have the slope. We find the slope by using the definition of the slope.

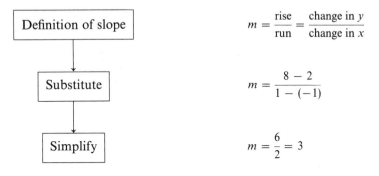

Now that know that slope m is 3, we can use the point-slope form. We use $(1, 8)$ as the given point, but $(-1, 2)$ would give us the same answer.

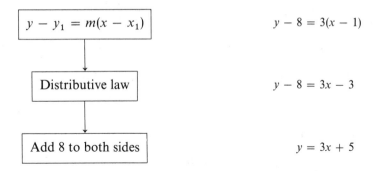

This is the equation of the line: $y = 3x + 5$. It is often helpful to check the answer by substituting both of the original points into this equation, $y = 3x + 5$.

Check $(-1, 2)$: $2 = 3(-1) + 5 = 2$ *Checks.*

Check $(1, 8)$: $8 = 3(1) + 5 = 8$ *Checks.*

Notice that the equations of the lines found in Examples 7 to 9 all had the final form

$$y = mx + b$$

Here m (the coefficient of x) is the slope. The term b is the y-intercept (see page 211). This is called the **slope-intercept form,** since the slope and y-intercept are so obvious. Figure 8.7 illustrates this.

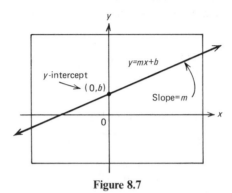

Figure 8.7

EXAMPLE 10   If an equation is in slope-intercept form, it is easy to determine the slope and the y-intercept.

(a) For $y = 2x + 7$, the slope is 2; the y-intercept is 7.

(b) For $y = -8x + 4$, the slope is -8; the y-intercept is 4.

(c) For $y = \dfrac{3}{4}x - \dfrac{5}{4}$, the slope is $\dfrac{3}{4}$; the y-intercept is $\dfrac{-5}{4}$.

(d) For $y = 6x$, the slope is 6; the y-intercept is 0.

(e) For $y = 7$, the slope is 0; the y-intercept is 7.

> To find the slope and y-intercept of a line:
>
> **1.** Put the equation into slope-intercept form, $y = mx + b$. (Solve for y.)
>
> **2.** The slope is m.
>
> **3.** The y-intercept is b.

EXAMPLE 11   Find the slope and y-intercept of $2x - 3y = 7$.

SOLUTION   We put this equation into slope-intercept form by solving for y.

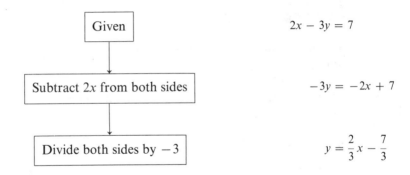

	$2x - 3y = 7$

Given

Subtract $2x$ from both sides $-3y = -2x + 7$

Divide both sides by -3 $y = \dfrac{2}{3}x - \dfrac{7}{3}$

Now we can see that the slope is $\dfrac{2}{3}$ and the y-intercept is $\dfrac{-7}{3}$.

EXAMPLE 12   Graph $y = 2x - 5$.

SOLUTION   In Chapter 6 we graphed this type of equation by finding various (x, y) pairs that satisfy the equation. Here we use the fact that the slope is 2 and the y-intercept is -5. We start at $(0, -5)$ and move along using the following:

If x changes 1, y changes 2

If x changes 2, y changes 4

If x changes 3, y changes 6

The graph appears as shown. (Review Examples 5 and 6.)

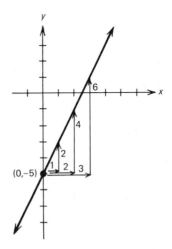

PROBLEM SET 8.2

Find the equation of each of the following lines through the given point and with the given slope.

See
Examples
7 and 8

1. $m = 2; (1, 5)$ **2.** $m = 5; (2, 9)$

3. $m = 7; (-2, 5)$ **4.** $m = 3; (-4, -5)$

5. $m = -3; (5, 2)$ **6.** $m = -1; (2, 10)$

7. $m = -10; (-6, 2)$ **8.** $m = -8; (-5, -7)$

9. $m = 0; (6, -1)$ **10.** $m = 0; (-4, 5)$

Find the equation of each of the lines through the given points.

See
Example 9

11. $(1, 5)$ and $(3, 9)$ **12.** $(2, 7)$ and $(4, 17)$

13. $(1, -1)$ and $(4, 11)$ **14.** $(6, 9)$ and $(4, 11)$

15. $(-5, 2)$ and $(-3, 10)$ **16.** $(-2, 10)$ and $(1, 1)$

17. $(3, 8)$ and $(7, 8)$ **18.** $(8, -1)$ and $(5, -1)$

For each of the following lines, determine the slope and y-intercept.

See
Example 10

19. $y = 7x + 3$ **20.** $y = 8x + 9$

21. $y = -3x + 4$ **22.** $y = -5x - 2$

23. $y = 6x - 8$ **24.** $y = \dfrac{-3}{2}x - \dfrac{1}{2}$

25. $y = 3x$ **26.** $y = x$

27. $y = 8$ **28.** $y = -6$

See
Example 11

29. $2x + y = 5$ **30.** $3x - y = 7$

31. $8x + 4y = 3$ **32.** $6x + 6y = 11$

33. $3x - 2y = 5$ **34.** $5x + 7y = 13$

35. $4x + 5y = 0$ **36.** $3x - 8y = 0$

Graph the following lines using only the slope and y-intercept.

See
Examples
6 and 12

37. $y = 3x - 2$ **38.** $y = 2x + 5$

39. $y = -x + 6$ **40.** $y = -4x + 2$

41. $y = \frac{1}{2}x + 3$ **42.** $y = \dfrac{-3}{2}x - 1$

Business applications **43.** A company finds that it can sell 5000 radios at $20 each and 6000 radios at $15 each. Find the equation of the line through (5000 radios, $20) and (6000 radios, $15). This is called the *demand line*.

44. A company buys a $50,000 truck. After 10 years, it is worth $5000. Find the equation of the line through (0 years, $50,000) and (10 years, $5000). This is called *straight-line depreciation*.

Consumer applications **45.** A family observes that it costs $1600 to drive their car 5000 miles per year. It costs them $2000 to drive their car 10,000 miles per year.
   (a) Find the equation of the line through (5000 miles, $1600) and (10,000 miles, $2000).
   (b) What is the slope? (This is the *cost per mile*.)
   (c) What is the y-intercept? (This is the *fixed cost*.)

46. A 6-ounce tube of Smile-eze Toothpaste costs 50 cents. A 12-ounce tube costs 80 cents.
   (a) Find the equation of the line through (6 ounces, 50 cents) and (12 ounces, 80 cents).
   (b) What is the y-intercept? (This is price of 0 ounces of toothpaste; in other words, it is the cost of the packaging.)
   (c) What is the slope? (This is the cost per ounce of the toothpaste.)

8.3   GRAPHING LINEAR INEQUALITIES

Recall that $ax + by = c$ is a linear equation. Its graph is a straight line. In this section we study **linear inequalities,** which have the standard forms

$$ax + by < c \qquad ax + by > c$$
$$ax + by \le c \qquad ax + by \ge c$$

Just as $x < 7$ is a half-line cut at $x = 7$ (see page 75), a linear inequality such as $2x + 5y < 10$ is a half-plane cut by the line $2x + 5y = 10$.

To graph a linear inequality:

1. Graph the corresponding line, $ax + by = c$.
2. Determine if a test point P (not on the line) satisfies the inequality. [Any point P will do, but $(0, 0)$ is usually a very convenient point if it is not on the line.]
3a. If P satisfies the inequality, shade the region *containing P*.
3b. If P does not satisfy the inequality, shade the region *not containing P*.
4a. If the inequality is $\le$ or $\ge$, include the line in the graph.
4b. If the inequality is $<$ or $>$, exclude the line from the graph (use a dashed line).

EXAMPLE 13   Graph $2x + 7y \geq 14$.

SOLUTION   We begin by graphing the line $2x + 7y = 14$. Recall from Section 6.3 that we do this by finding the x- and y-intercepts.

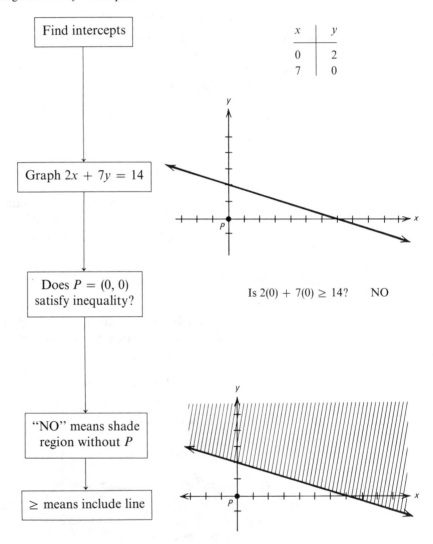

x	y
0	2
7	0

Is $2(0) + 7(0) \geq 14$?   NO

EXAMPLE 14   Graph $5x + 4y < 40$.

SOLUTION   We begin by graphing the line $5x + 4y = 40$.

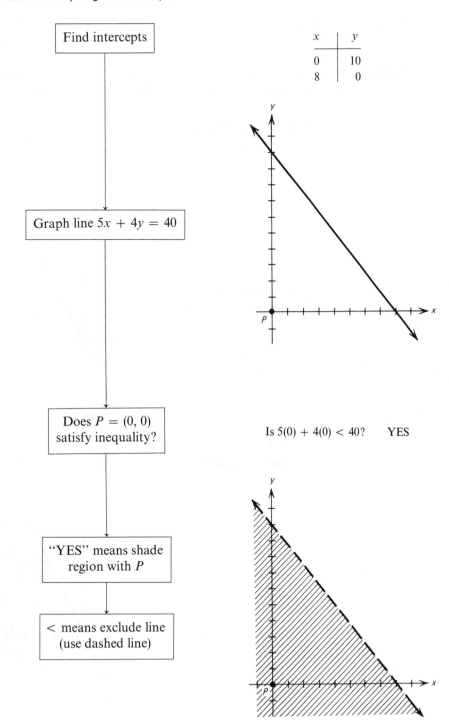

Find intercepts

x	y
0	10
8	0

Graph line $5x + 4y = 40$

Does $P = (0, 0)$ satisfy inequality?

Is $5(0) + 4(0) < 40$?   YES

"YES" means shade region with P

$<$ means exclude line (use dashed line)

EXAMPLE 15   Graph $y \leq 2x$.

SOLUTION   We first graph the line $y = 2x$. Since the point $(0, 0)$ is on this line, we must choose another test point. Here we use $P = (3, 2)$, which was chosen randomly.

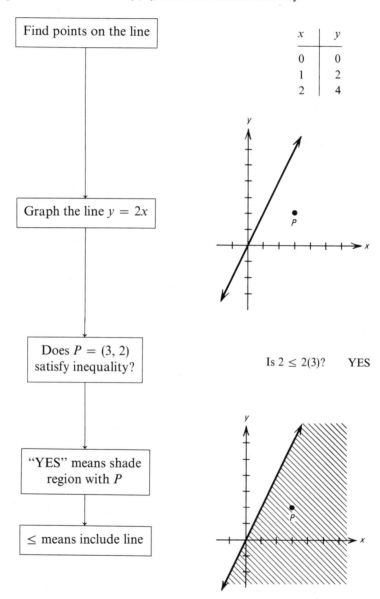

x	y
0	0
1	2
2	4

Find points on the line

Graph the line $y = 2x$

Does $P = (3, 2)$ satisfy inequality?

Is $2 \leq 2(3)$?   YES

"YES" means shade region with P

$\leq$ means include line

PROBLEM SET 8.3

Graph the following inequalities.

See   **1.** $x + y \leq 2$ **2.** $x - y > 5$
Examples
13 and 14 **3.** $x + 2y \geq 4$ **4.** $3x + y < 6$

        5. $2x - 3y > 6$ **6.** $4x + 3y \leq 12$

        7. $5x + 2y < 20$ **8.** $2x - 9y \geq 18$

        9. $4x - 5y \leq 20$ **10.** $4x + 7y > 28$

See   **11.** $y \geq 2x + 3$ **12.** $y < 4x - 2$
Example 15 **13.** $y > x - 5$ **14.** $y \leq 6 - x$

      15. $y < 3x$ **16.** $y \geq -x$

      17. $y \leq -4x$ **18.** $y > 5x$

Health   **19.** A patient needs 56 grams of protein. A glass of milk provides 8 grams
application   of protein, and a slice of bread provides 2 grams of protein. The protein
          relation between milk x and bread y is

$$8x + 2y \geq 56$$

Graph this inequality.

Business   **20.** An automobile company manufactures two cars, Beavers and Raccoons.
application   The number of Beavers x and Raccoons y it produces is related to the
          amount of available steel by the relation

$$200x + 250y \leq 600{,}000$$

Graph this inequality.

Environmental **21.** A plant manufactures two products. Product 1 (x) produces 15 grams of
application SO_2 per item, and product 2 (y) produces 25 grams of SO_2 per item. The
          Environmental Protection Agency allows the manufacturer a maximum
          of 75,000 grams of SO_2 per day. Graph the inequality that this situation
          produces.

$$15x + 25y \leq 75{,}000$$

8.4   GRAPHING QUADRATRIC EQUATIONS

Thus far we have considered only graphs of linear equations and inequalities. In this section we consider the graph of the **quadratic equation** in two variables. This has the standard form

$$y = ax^2 + bx + c \quad (a \neq 0)$$

Figure 8.8 illustrates the graphs of two typical quadratic equations. The shape of this curve is called a **parabola.** The parabola is seen in science as the path of a thrown object or the cross section of radar scanners or flashlight reflectors.

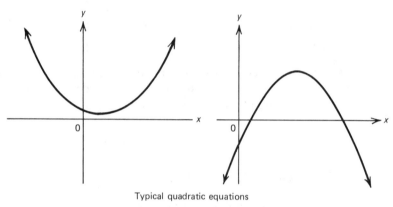

Typical quadratic equations

Figure 8.8

To graph a quadratic equation:

1. Choose an appropriate set of x-values to use.

2. Make a table of (x, y) pairs by substituting the x-values into the equation for y.

3 Plot these points on a graph.

4. Draw a smooth curve through these points.

EXAMPLE 16   Graph $y = x^2$ for $-3 \leq x \leq 3$.

SOLUTION   We substitute the x-values, -3 to 3, into the equation $y = x^2$.

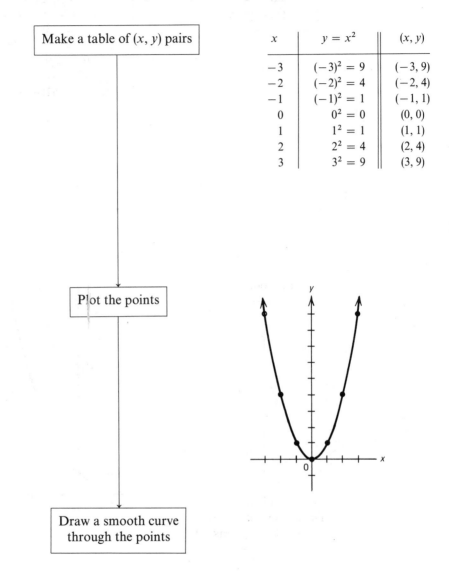

| | Make a table of (x, y) pairs |

x	$y = x^2$	(x, y)
-3	$(-3)^2 = 9$	$(-3, 9)$
-2	$(-2)^2 = 4$	$(-2, 4)$
-1	$(-1)^2 = 1$	$(-1, 1)$
0	$0^2 = 0$	$(0, 0)$
1	$1^2 = 1$	$(1, 1)$
2	$2^2 = 4$	$(2, 4)$
3	$3^2 = 9$	$(3, 9)$

Plot the points

Draw a smooth curve
through the points

EXAMPLE 17   Graph $y = x^2 + 2x + 3$ for $-2 \leq x \leq 4$.

SOLUTION   We make a table of (x, y) pairs for the given values, -2 to 4.

Make a table of (x, y) pairs

x	$y = x^2 + 2x + 3$	(x, y)
-2	$(-2)^2 + 2(-2) + 3 = 3$	$(-2, 3)$
-1	$(-1)^2 + 2(-1) + 3 = 2$	$(-1, 2)$
0	$0^2 + 2(0) + 3 = 3$	$(0, 3)$
1	$1^2 + 2(1) + 3 = 6$	$(1, 6)$
2	$2^2 + 2(2) + 3 = 11$	$(2, 11)$
3	$3^2 + 2(3) + 3 = 18$	$(3, 18)$
4	$4^2 + 2(4) + 3 = 27$	$(4, 27)$

Plot points

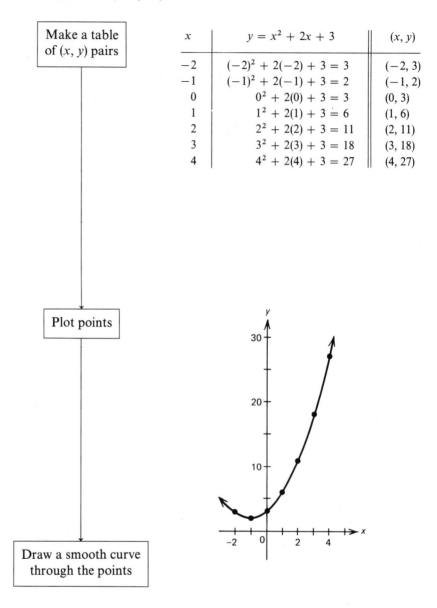

Draw a smooth curve through the points

EXAMPLE 18   Graph $y = -2x^2 + 7x + 1$ for $-1 \le x \le 5$.

SOLUTION   We start by making a table using the x-values, -1 to 5.

Make a table of (x, y) pairs

x	$y = -2x^2 + 7x + 1$	(x, y)
-1	$-2(-1)^2 + 7(-1) + 1 = -8$	$(-1, -8)$
0	$-2(0)^2 + 7(0) + 1 = 1$	$(0, 1)$
1	$-2(1)^2 + 7(1) + 1 = 6$	$(1, 6)$
2	$-2(2)^2 + 7(2) + 1 = 7$	$(2, 7)$
3	$-2(3)^2 + 7(3) + 1 = 4$	$(3, 4)$
4	$-2(4)^2 + 7(4) + 1 = -3$	$(4, -3)$
5	$-2(5)^2 + 7(5) + 1 = -14$	$(5, -14)$

Plot points

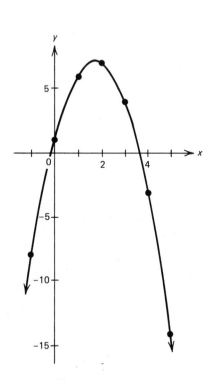

Draw a smooth curve through the points

PROBLEM SET 8.4

Graph the following quadratic equations for the indicated x-values.

See
Examples
16, 17, and 18

1. $y = x^2$ $0 \leq x \leq 6$

2. $y = x^2 + 2$ $-3 \leq x \leq 3$

3. $y = 3x^2$ $-2 \leq x \leq 4$

4. $y = x^2 - 5$ $-1 \leq x \leq 5$

5. $y = 2x^2 + 1$ $0 \leq x \leq 6$

6. $y = -x^2 + 5$ $-1 \leq x \leq 5$

7. $y = x^2 + x + 1$ $-2 \leq x \leq 4$

8. $y = x^2 - x + 2$ $-3 \leq x \leq 3$

9. $y = x^2 - 3x - 5$ $-2 \leq x \leq 4$

10. $y = x^2 + 4x + 8$ $-1 \leq x \leq 5$

11. $y = -x^2 + x + 3$ $0 \leq x \leq 6$

12. $y = -x^2 - x - 8$ $-1 \leq x \leq 5$

13. $y = -2x^2 - x + 5$ $-2 \leq x \leq 4$

14. $y = -3x^2 + 6x + 1$ $-3 \leq x \leq 3$

Business
application

15. The profit on an item is given by

$$P = -x^2 + 200x - 1000$$

where x is the number of items produced.
(a) Graph this equation. (Use $x = 0, 25, 50, 75, 100, 125,$ and 150.)
(b) What number of items produces a maximum profit?

Science
application

16. The height of an object thrown into the air is given by

$$h = -16t^2 + 80t + 10$$

(a) Graph this equation for $0 \leq t \leq 5$ (t is the time in seconds).
(b) At what time is the height maximum?

Life science
application

17. The population of a bacteria culture is given by

$$P = -100t^2 + 600t + 4000$$

where t is the time in hours.
(a) Graph this equation for $0 \leq t \leq 6$.
(b) At what time is the population maximum?

CHAPTER 8 SUMMARY

Important Words and Phrases

equation of a line (8.2)
linear inequalities (8.3)
parabola (8.4)
point-slope form (8.2)

quadratic equation in two
   variables (8.4)
slope (8.1)
slope-intercept form (8.2)
y-intercept (8.2)

Important Definitions and Theorems

The slope of the line through two points (x_1, y_1) and (x_2, y_2) is given by

$$m = \frac{\text{rise}}{\text{run}} = \frac{\text{change in } y}{\text{change in } x} = \frac{y_2 - y_1}{x_2 - x_1}$$

The equation of the line through (x_1, y_1) with slope m is given by

$$y - y_1 = m(x - x_1)$$

Important Procedures

To find the slope and y-intercept of a line:
1. Put the equation into slope-intercept form, $y = mx + b$.
2. The slope is m; the y-intercept is b.

To graph a linear inequality:
  1. Graph the corresponding equation, $ax + by = c$.
  2. Determine if a point P (not on the line) satisfies the inequality.
3a. If P satisfies the inequality, shade the region with P.
3b. If P does not satisfy the inequality, shade the region without P.
4a. If the inequality is $\leq$ or $\geq$, include the line.
4b. If the inequality is $<$ or $>$, exclude the line (use a dashed line).

To graph a quadratic equation:
1. Make a table of (x, y) pairs by substituting the various x-values into the equation for y.
2. Plot the points on a graph.
3. Draw a smooth curve through the points.

CHAPTER 8 REVIEW EXERCISES

Find the slopes of the lines through the following pairs of points.

1. (1, 7) and (3, 15)
2. (2, 5) and (5, 2)

3. (3, −8) and (5, −5)
4. (−2, 1) and (5, −1)

Graph the following lines.

5. $m = 2$; through (1, 2)
6. $m = -3$; through (−2, 1)

7. $y = 4x + 1$
8. $y = -x + 7$

Find the equations of the following lines.

9. $m = 3$; through (4, −2)
10. $m = -5$; through (−1, −8)

11. Through (1, 3) and (4, 15)
12. Through (2, −6) and (4, −12)

Graph the following.

13. $3x - 4y > 12$
14. $2x + 5y \le 20$

15. $y \le 2x + 1$
16. $y < -x$

17. $y = x^2$ $-2 \le x \le 4$

18. $y = -2x^2 + 1$ $-3 \le x \le 3$

19. $y = 3x^2 - x + 1$ $-1 \le x \le 5$

ANSWERS TO
SELECTED EXERCISES

PROBLEM SET 1.1

1. True **3.** False **5.** True **7.** False **9.** False **11.** True **13.** 9 **15.** 7
17. 14 **19.** 5 **21.** 9 **23.** 27 **25.** $a = $ b **27.** $x \neq 5$ **29.** $m < n$ **31.** $u \geq v$

33. $x + 10$ **35.** $a - b$ **37.** hk **39.** xy **41.** $\dfrac{p}{q}$ **43.** u^2 **45.** a^3 **47.** $a = b + 4$

49. 13 **51.** 4 **53.** 4 **55.** 4 **57.** 15 **59.** 64 **61.** 3 **63.** 2 **65.** 2 **67.** 38
69. 28 **71.** 1 **73.** \$410,000 **75.** 0.488

PROBLEM SET 1.2

1. 5 **3.** 20 **5.** 4 **7.** 9 **9.** 9 **11.** 36 **13.** 6 **15.** 11 **17.** 3 **19.** 7
21. 8 **23.** 19 **25.** 29 **27.** 9 **29.** 4 **31.** 5 **33.** 3 **35.** 3, 4, and 5
37. 2 and 3 **39.** 4 and 5 **41.** 1, 2, and 3 **43.** 1, 2, and 3 **45.** **(a)** $10°$; **(b)** $25°$
47. 100 milligrams

PROBLEM SET 1.3

1. 1 **3.** 0 **5.** -5 **7, 9, 11.** **13.** 55 **15.** 20 **17.** -4

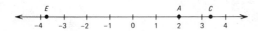

19. -200 **21.** 4000 **23.** $18; -18; 18$ **25.** $-11; 11; 11$ **27.** $2; -2; 2$ **29.** $-9; 9; 9$
31. 6 **33.** 12 **35.** -6 **37.** False **39.** False **41.** False

43.

PROBLEM SET 1.4

1. **3.** **5.**

7. -7 **9.** 9 **11.** 9 **13.** -1 **15.** 18 **17.** -16 **19.** -35 **21.** 36 **23.** -18
25. 11 **27.** -3 **29.** -18 **31.** -9 **33.** 7 **35.** 7 **37.** 3 **39.** 4 **41.** -4
43. -21 **45.** -1 **47.** -13 **49.** -8 **51.** -3 **53.** -22

PROBLEM SET 1.5

1. 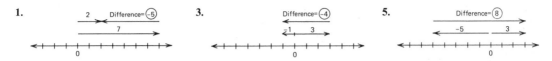   **3.**   **5.**

7. -7 **9.** -8 **11.** 5 **13.** -6 **15.** 3 **17.** 5 **19.** -6 **21.** -17 **23.** 6
25. -4 **27.** -8 **29.** -16 **31.** -12 **33.** -1 **35.** -7 **37.** 12 **39.** -10
41. -22 **43.** -1 **45.** 6 **47.** 12 **49.** 6 **51.** $5000; -8000; 5000; -10{,}000$

PROBLEM SET 1.6

1. 14 **3.** -24 **5.** -20 **7.** 30 **9.** 40 **11.** -48 **13.** 44 **15.** 50 **17.** $\dfrac{-1}{12}$

19. 0 **21.** 8 **23.** 0 **25.** 10 **27.** -1 **29.** 20 **31.** 22 **33.** -14 **35.** -70
37. -24 **39.** -11 **41.** 6 **43.** -20 **45.** -18 **47.** $3; -3; -6; 8$

PROBLEM SET 1.7

1. -2 **3.** 5 **5.** 2 **7.** -4 **9.** 2 **11.** -6 **13.** Undefined (no such number)
15. 30 **17.** 0 **19.** 1 **21.** 10 **23.** -2 **25.** -2 **27.** 4 **29.** 2 **31.** -2

33. -2 **35.** -3 **37.** -1 **39.** $\dfrac{1}{8}$ **41.** $\dfrac{-3}{2}$ **43.** None   **45.** -28 **47.** $\dfrac{-22}{15}$

49. $\dfrac{-2}{99}$ **51.** $2; \dfrac{1}{2}; \dfrac{1}{10}; 5$ **53.** $2; -1; -3; 0$

PROBLEM SET 1.8

1. Q, R **3.** H, R **5.** Q, R **7.** H, R **9.** I, Q, R **11.** Q, R **13.** False   **15.** False
17. True   **19.** True   **21.** Identity   **23.** Commutative   **25.** Distributive   **27.** Commutative
29. Distributive

CHAPTER 1 REVIEW EXERCISES

1. True   **2.** False   **3.** True   **4.** $x \le 10$ **5.** $5t$ **6.** $a = b + 8$ **7.** 4 **8.** $6, 7$

9. $-3; \dfrac{1}{3}; 3$ **10.** $7; \dfrac{-1}{7}; 7$ **11.** $\dfrac{2}{5}; \dfrac{-5}{2}; \dfrac{2}{5}$ **12.** 4 **13.** -2 **14.** 25 **15.** 8 **16.** 2

17. 27 **18.** -2 **19.** -3 **20.** 1 **21.** -10 **22.** -35 **23.** 12 **24.** 4 **25.** 3

26. -5 **27.** Meaningless   **28.** 2 **29.** $\dfrac{14}{15}$ **30.** -4 **31.** -11 **32.** 12 **33.** -8

34. 3 **35.** 4 **36.** Distributive   **37.** Identity   **38.** Inverse

PROBLEM SET 2.1

1. $4x + 20$ **3.** $10 - 5x$ **5.** $-12 + 6x - 3y$ **7.** $7a + 8$ **9.** $6r - 7$ **11.** $4p + 1$
13. $9x - 2$ **15.** $7a - 12$ **17.** $6g - 5$ **19.** $9x + 10$ **21.** $6a - 10$ **23.** $-2u - 7$
25. $3x - 16$ **27.** $8k - 12$ **29.** $-2m - 12$ **31.** $2x + 7$ **33.** $14x - 12$ **35.** $8u - 7$
37. $11x - 26$ **39.** $-3m - 40$

PROBLEM SET 2.2

1. Equivalent   **3.** Not equivalent   **5.** 15   **7.** 10   **9.** 1   **11.** 4   **13.** -3 **15.** -14
17. 9   **19.** 3   **21.** -2 **23.** 6   **25.** 9   **27.** -8 **29.** \$120,000

PROBLEM SET 2.3

1. 5   **3.** 4   **5.** -7 **7.** -8 **9.** 1   **11.** 12   **13.** 60   **15.** -16 **17.** 6

19. -4 **21.** 7   **23.** $\dfrac{-3}{5}$ **25.** 15   **27.** 36   **29.** -21 **31.** 1   **33.** -1 **35.** -3

37. 2   **39.** 1   **41.** 4   **43.** 1   **45.** 150   **47.** \$5000   **49.** $20°$

PROBLEM SET 2.4

1. 5   **3.** -5 **5.** -3 **7.** -1 **9.** 3   **11.** 5   **13.** 10   **15.** -4 **17.** 5   **19.** 4
21. 2   **23.** 2   **25.** 4   **27.** -5 **29.** -3 **31.** 5   **33.** 6   **35.** 4

PROBLEM SET 2.5

1. 17 and 25   **3.** 7 and 35   **5.** 23 and 24   **7.** 40   **9.** 15%   **11.** 820   **13.** 38 and 76
15. 425 and 575   **17.** 23   **19.** 17 and 11

PROBLEM SET 2.6

1. $x < 4$ **3.** $k \geq 11$ **5.** $m > 5$

7. $x \leq 7$ **9.** $a < 33$ **11.** $y \geq 5$

13. $x > 10$ **15.** $k \geq -3$ **17.** $m < -2$

19. $y \leq 2$ **21.** $x > 6$ **23.** $x \geq -40$

25. $t < -20$ **27.** $f \leq 6$ **29.** $7 > x$

31. $a \leq -2$ **33.** $y < -1$ **35.** $x \geq -6$

37. $x > 2$ **39.** $x \leq -1$ **41.** $x < 1$

43. $-10 > a$ **45.** $y \leq 70$

CHAPTER 2 REVIEW EXERCISES

1. $21x + 14$ **2.** $5a - 18$ **3.** $-4u + 2$ **4.** $5t - 13$ **5.** $-23t + 12$ **6.** Yes   **7.** 13
8. -3 **9.** 11   **10.** 8   **11.** -6 **12.** 20   **13.** -90 **14.** -8 **15.** 2   **16.** -4

17. 4   **18.** -3 **19.** 2   **20.** 3   **21.** $x < 13$

22. $a \le 10$ **23.** $k \ge -5$ **24.** $x < -63$

25. $x < 10$ **26.** $x \le -3$ **27.** 8 and 12   **28.** 1500

29. 9 and 11   **30.** 6 and 15

PROBLEM SET 3.1

1. x^6 **3.** 2^8 **5.** $(p + q)^2$ **7.** 16   **9.** 1000   **11.** -64 **13.** 10,000   **15.** 0.36

17. $\dfrac{4}{25}$ **19.** x^5 **21.** 3^9 **23.** $(a + b)^7$ **25.** a^{12} **27.** 10^9 **29.** x^{10} **31.** z^{16}

33. 8^{21} **35.** 9^8 **37.** $27x^3$ **39.** $16a^2b^2$ **41.** $81a^4b^4c^4$ **43.** $125a^3b^6$ **45.** 2^{46}
47. $9 \cdot 10^{16}$

PROBLEM SET 3.2

1. x^4 **3.** t^4 **5.** 1   **7.** 1   **9.** m^2 **11.** $\dfrac{1}{3^6}$ **13.** $\dfrac{9}{a^2}$ **15.** $\dfrac{m^3}{8}$ **17.** $\dfrac{16r^4}{p^4}$

19. $\dfrac{x^3y^3}{z^3}$ **21.** $\dfrac{x^8y^4}{z^4}$ **23.** $\dfrac{8m^3k^6}{125n^9}$ **25.** $\dfrac{1}{x^6}$ **27.** $\dfrac{1}{32}$ **29.** $\dfrac{1}{25}$ **31.** $\dfrac{1}{10,000}$ **33.** 1

35. 1   **37.** $\dfrac{1000}{(1.09)^8}$ **39.** $\dfrac{1}{10,000,000}$

PROBLEM SET 3.3

1. 5   **3.** 10   **5.** -1 **7.** $2x$ **9.** $-3pq$ **11.** $3xy$ **13.** $-8abcd$ **15.** $12x^3$
17. $-24r^5$ **19.** $15a^3b^3$ **21.** $-24x^3y^5z^5$ **23.** $60u^{10}v^8$ **25.** $-20a^3b^6c^4$ **27.** $2x^3$

29. $\dfrac{-4}{u}$ **31.** $\dfrac{-12b}{a}$ **33.** $-2uv^2$ **35.** $\dfrac{yw^6}{2x^3}$ **37.** $\dfrac{3n^2k}{m^2j^5}$

PROBLEM SET 3.4

1. 2; trinomial   **3.** 3; binomial   **5.** 4; polynomial   **7.** 1; monomial   **9.** $7x^2 + 12x + 2$
11. $9u^4 + 3u^3 + 3u^2 + 6u + 4$ **13.** $13t^4 + 5t^3 + 7t^2 + 6t - 3$ **15.** $9x^2 - 2x + 4$
17. $11r^3 - r + 4$ **19.** $16t^3 + 16t^2 - 9t - 15$ **21.** $x^2 - 8x + 3$ **23.** $4m^3 + 7m^2 - 13m + 7$
25. $5t^3 - 8t^2 - 9t - 1$ **27.** $3x^4 - x^3 - 5x^2 + 2x - 4$ **29.** $7x^2 - 2x$ **31.** $-x + 21$
33. $6r^2 + r - 3$ **35.** $6m^3 + 7m^2 - 11m + 1$ **37.** $-x^2 + 75x - 2000; -0.5x^2 + 37x - 8000$

PROBLEM SET 3.5

1. $15x^3 + 6x^2 - 18x$ **3.** $4a^3b + 8a^2b^2 + 4ab^3$ **5.** $56p^2q^2 - 14pq - 28p^4$
7. $3s^3t - 12s^3t^3 + 18s^5t^2$ **9.** $6x^2 + 13x - 5$ **11.** $4u^2 + 25u - 21$ **13.** $8t^2 + 32t + 30$
15. $32k^2 - 44k + 5$ **17.** $2a^2 - 7ab + 3b^2$ **19.** $20m^2 + 9mn - 18n^2$ **21.** $12r^2 + 5rs - 2s^2$
23. $x^3 + 3x^2 - 13x - 15$ **25.** $2u^3 + 13u^2 + 13u - 3$ **27.** $a^3 + a^2b - ab^2 + 2b^3$
29. $8r^3 + 2r^2t - 3rt^2 + 18t^3$ **31.** $6x^5 - 5x^4 + 26x^3 - 14x^2 + 27x - 10$

33. $30u^6 - 21u^5 + 46u^4 - 24u^3 + 32u^2 - 9u + 9$ **35.** $2m^6 + 2m^5 - 15m^4 + 34m^3 - 31m^2 + 14m - 2$
37. $x^3 + 6x^2 + 11x + 6$ **39.** $u^3 - 6u^2 - u + 30$ **41.** $x^3 - 5x^2 - 13x - 7$
43. $2t^4 - 13t^3 + 11t^2 + 19t + 5$ **45.** $11x^3 + (-22L - 21)x^2 + (11L^2 + 42L)x - 21L^2$

PROBLEM SET 3.6

1. $x^2 + 5x + 6$ **3.** $u^2 + 4u - 21$ **5.** $2a^2 + 15a + 7$ **7.** $2z^2 + 9z - 56$ **9.** $6u^2 + 5u - 6$
11. $24k^2 + 6k - 3$ **13.** $21a^2 - 29ab - 10b^2$ **15.** $35u^2 + 8uv - 3v^2$ **17.** $6m^2 - 13mn - 5n^2$
19. $x^2 + 8x + 16$ **21.** $r^2 - 16r + 64$ **23.** $4c^2 - 20c + 25$ **25.** $25x^2 - 30xy + 9y^2$
27. $9u^2 - 48uv + 64v^2$ **29.** $81p^2 + 180pq + 100q^2$ **31.** $r^2 - s^2$ **33.** $u^2 - 4$ **35.** $9 - a^2$
37. $4x^2 - 9$ **39.** $36a^2 - 16b^2$ **41.** $81x^4y^4 - 4$ **43.** $45,000 + 30,000t + 5000t^2$

PROBLEM SET 3.7

1. $4a + 2$ **3.** $2a^2 - 1$ **5.** $4b - 3a^3 + 5a^2$ **7.** $9bc - 3a^3 - 5a^2$ **9.** $2x^2y^4 - 3x^5$

11. $xy^2 - 10x^5y^2$ **13.** $2x - 3y + \dfrac{4y^2}{x}$ **15.** $8x^2y^4 - 6 + \dfrac{2}{xy^2}$ **17.** $2a - 3b^3$

19. $11u^6 - 4uv^5 - 3u$ **21.** $2mn - 3m^3n$ **23.** $a^4 + a^2 + \dfrac{1}{a}$ **25.** $\dfrac{11}{st} - \dfrac{6}{t^2} + \dfrac{8s}{t}$

27. $\dfrac{7a}{bc} - \dfrac{5a^2}{c^2} + \dfrac{2a^3}{b}$ **29.** (a) $-16t^2 - 32th - 16h^2 + 120$ (b) $-32th - 16h^2$ (c) $-32t - 16h$

PROBLEM SET 3.8

1. $18\frac{1}{29}$ **3.** $22\frac{51}{53}$ **5.** $141\frac{45}{46}$ **7.** $x + 7$ **9.** $x - 8$ **11.** $2x + 3$ **13.** $2t - 7$

15. $5m^2 - 3m + 1$ **17.** $8u^2 - u + 4$ **19.** $x + 9 + \dfrac{29}{x - 4}$ **21.** $w - 14 + \dfrac{107}{w + 8}$

23. $3k + 9 + \dfrac{52}{k - 5}$ **25.** $2x^2 + 3x - 4 + \dfrac{-5}{2x - 5}$ **27.** $4u^2 - 2u - 3 + \dfrac{-4}{2u - 7}$

29. $4m^2 - m + 3 + \dfrac{13}{5m - 4}$ **31.** $x^2 + x + 1$ **33.** $x^3 - x^2 + x - 1 + \dfrac{2}{x + 1}$

35. $u^3 + 2 + \dfrac{4}{u^3 - 1}$

CHAPTER 3 REVIEW EXERCISES

1. t^5 **2.** 8^3 **3.** 32 **4.** 100 **5.** x^{11} **6.** a^{20} **7.** $27u^3v^6$ **8.** m^2 **9.** $\dfrac{1}{n^5}$

10. $\dfrac{16}{k^4}$ **11.** $\dfrac{16a^4b^2}{c^6}$ **12.** $\dfrac{1}{32}$ **13.** 1 **14.** 6 **15.** -1 **16.** 2; trinomial

17. 4; binomial   **18.** $-5pq$ **19.** $-24x^5y^8$ **20.** $\dfrac{-4c^5}{a}$ **21.** $8t^3 + 5t^2 - 3t + 5$

22. $9u^4 + 3u^3 - 7u^2 - 9u + 7$ **23.** $-x^3 - 12x^2 + 3x + 3$ **24.** $5m^2 + 2$ **25.** $30a^6 - 18a^4 - 6a^3$
26. $24u^2 - 31u - 15$ **27.** $2x^3 + 13x^2 + 27x + 28$ **28.** $x^3 - 2x^2 - 31x - 28$
29. $2t^5 - 3t^4 - 6t^3 + 9t^2 + 2t - 3$ **30.** $24r^2 + 18r - 15$ **31.** $4x^2 - 20x + 25$

32. $9a^2 + 12a + 4$ **33.** $49p^2 - 25q^2$ **34.** $3x - 6x^2y^2 + 9x^3y^3$ **35.** $2b - \dfrac{3b^2}{a} + \dfrac{4a^5}{b}$

36. $x + 8$ **37.** $2x^2 - 3x + 4$ **38.** $3u^2 + 5u - 2 + \dfrac{6}{2u + 5}$ **39.** $a^4 + a^3 + a^2 + a + 1$

PROBLEM SET 4.1

1. $3 \cdot 7$ **3.** $3 \cdot 3$ **5.** $2 \cdot 3 \cdot 7$ **7.** $5 \cdot 5$ **9.** $3 \cdot 3 \cdot 3$ **11.** $2 \cdot 2 \cdot 5 \cdot 5$ **13.** $3 \cdot 5 \cdot 7$
15. $2 \cdot 2 \cdot 2 \cdot 2 \cdot 3$ **17.** $2 \cdot 2 \cdot 2 \cdot 5 \cdot 5$ **19.** $2 \cdot 3 \cdot 3 \cdot 3 \cdot 5$ **21.** Prime **23.** $3 \cdot 7 \cdot 7$ **25.** $5 \cdot 17$
27. $5 \cdot 5 \cdot 7$ **29.** $3 \cdot 3 \cdot 11$

PROBLEM SET 4.2

1. 4 **3.** 6 **5.** 5 **7.** $3x$ **9.** ab **11.** mn^4 **13.** $9p^2q^4$ **15.** $2(x + 3y)$
17. $8(2r - s)$ **19.** $5(5x + 7y)$ **21.** $4x(2 - x)$ **23.** $3t^2(2t - 3)$ **25.** $2k^2(1 - 2k + 4k^2)$
27. $2u^3(5 - 6u^4 + 12u^{13})$ **29.** $a^3b^3(a + b)$ **31.** $2x^2y(y^3 - 3x)$ **33.** $5u^2v(2u^2 - 3v + 4uv^3)$
35. $6x^3y(3x^4y - 4y^4 + 5x)$ **37.** $u^2vw^2(v^2w^2 + u^3)$ **39.** $4k^2mn^7(2k^2m^2n^3 - 3)$
41. $4xyz(y - 2x - 3z)$ **43.** $P(1 + r)$

PROBLEM SET 4.3

1. $(x + 1)(x + 3)$ **3.** $(a + 2)(a + 6)$ **5.** $(r + 5)(r + 3)$ **7.** $(p + 6)(p - 3)$ **9.** $(t - 2)(t - 5)$
11. Does not factor with integers **13.** $(b - 6)(b - 2)$ **15.** $(x + 1)(x + 7)$ **17.** $(d + 8)(d - 3)$
19. $(y + 5)(y - 4)$ **21.** $(s - 1)(s + 3)$ **23.** $(r + 5)(r - 3)$ **25.** $(k - 6)(k - 4)$
27. $2(x - 5)(x + 1)$ **29.** $10(r + 2)(r + 4)$ **31.** $-2(x - 7)(x + 3)$ **33.** $t(t + 9)(t - 5)$
35. $2r(r - 1)(r + 6)$ **37.** $6z^3(z - 1)(z + 2)$ **39.** $4y^4(y - 6)(y + 1)$ **41.** $-16(t - 2)(t - 3)$

PROBLEM SET 4.4

1. $(2x + 3)(x + 1)$ **3.** $(2u + 1)(u + 5)$ **5.** $(4y - 3)(2y + 1)$ **7.** $(3x + 2)(x - 7)$
9. $(5k - 2)(k + 6)$ **11.** $(5y + 1)(2y - 3)$ **13.** Does not factor with integers **15.** $(5a + 4)(a - 7)$
17. $(3q - 5)(2q - 3)$ **19.** $(6z + 5)(z + 3)$ **21.** $(2p - 5)(4p - 3)$ **23.** $(x - 5)(10x - 3)$
25. $(8k + 3)(k - 7)$ **27.** $(5r + 3)(6r + 5)$ **29.** $(5t + 3)(7t - 2)$ **31.** $2a(3a + 2)(2a + 1)$
33. $10t^2(4t + 1)(3t - 2)$ **35.** $3u(5u + 4)(2u - 1)$ **37.** $2k^2(4k - 3)(3k - 2)$ **39.** $y^2(5y + 2)(y + 8)$
41. $-a^3(3a - 4)(7a + 3)$ **43.** $2(2x - 3)(4x - 5)$

PROBLEM SET 4.5

1. $(x + 1)(x - 1)$ **3.** $(p + 7)(p - 7)$ **5.** $(t - 2)(t + 2)$ **7.** $(m - 11)(m + 11)$
9. $(2x + 1)(2x - 1)$ **11.** $(4p - 5)(4p + 5)$ **13.** $(9t - 8)(9t + 8)$ **15.** $(3y - 5)(3y + 5)$
17. $(7k - 11)(7k + 11)$ **19.** $(2a - 9)(2a + 9)$ **21.** $(x^2 + 1)(x - 1)(x + 1)$
23. $(4u^2 + 1)(2u - 1)(2u + 1)$ **25.** $(16 + y^2)(4 - y)(4 + y)$ **27.** $(x + 1)^2$ **29.** $(b - 9)^2$
31. $(y + 7)^2$ **33.** $(2x + 5)^2$ **35.** $(3a - 7)^2$ **37.** $(4u + 3)^2$ **39.** $(6m - n)^2$ **41.** $5000(t + 3)^2$

PROBLEM SET 4.6

1. $-1, 2$ **3.** $-5, -6$ **5.** $-7, -4$ **7.** $\dfrac{-9}{2}, \dfrac{-2}{3}$ **9.** $0, \dfrac{-5}{8}, \dfrac{4}{5}$ **11.** $-1, -3$ **13.** $2, -4$

15. $7, -3$ **17.** $9, -5$ **19.** $\dfrac{-1}{2}, -5$ **21.** $7, \dfrac{-3}{2}$ **23.** $\dfrac{6}{5}, \dfrac{1}{3}$ **25.** $\dfrac{2}{5}, -7$ **27.** $-5, -2$

29. $4, -3$ **31.** $-1, -7$ **33.** $\dfrac{-1}{2}, 5$ **35.** $\dfrac{5}{2}, \dfrac{3}{2}$ **37.** $\dfrac{1}{4}, -7$ **39.** $-1, 0, 1$ **41.** $0, -2, 2$

43. $0, \dfrac{-2}{3}, \dfrac{2}{3}$ **45.** $0, -4, -5$ **47.** $4, 25$ **49.** $1, 3$

PROBLEM SET 4.7

1. $4, 8$ **3.** $2, 9$ **5.** $8, 9$ **7.** $11, 13$ **9.** $1, 11$ **11.** $8, 9$ **13.** 7 **15.** 4 by 10
17. 4 by 8

CHAPTER 4 REVIEW EXERCISES

1. Prime **2.** $3 \cdot 3 \cdot 3 \cdot 3$ **3.** $7 \cdot 13$ **4.** 2 **5.** $2mn^3$ **6.** $5(2x - 3y)$ **7.** $2x^2(2 - 3x^3)$
8. $6ab^3(ab - 2a^3 + 4b^2)$ **9.** $5u^2v^3w^3t^3(5wt^2 - 7u)$ **10.** $(x + 2)(x + 9)$ **11.** $(y - 8)(y + 3)$
12. $(k - 10)(k - 1)$ **13.** $2a(a - 7)(a + 4)$ **14.** $(2x + 3)(x + 5)$ **15.** $(4r - 3)(3r + 1)$
16. $(3u - 2)(2u - 5)$ **17.** $p^2(4p - 3)(p + 5)$ **18.** $(x - 8)(x + 8)$ **19.** $(2 - 9p)(2 + 9p)$
20. $(4q^2 + 9)(2q + 3)(2q - 3)$ **21.** $(k + 2)^2$ **22.** $(3m - 5)^2$ **23.** $-7, 4$ **24.** $8, -1$

25. $\dfrac{-5}{3}, \dfrac{1}{2}$ **26.** $5, -7$ **27.** $0, -5, 5$ **28.** $10, 12$ **29.** $2, 9$ **30.** 5 by 11

PROBLEM SET 5.1

1. $\dfrac{3}{10}$ **3.** $\dfrac{5}{14}$ **5.** $\dfrac{4}{5}$ **7.** $\dfrac{7}{13}$ **9.** $\dfrac{15}{16}$ **11.** $\dfrac{3}{2}$ **13.** $\dfrac{12}{11}$ **15.** $\{\text{reals}, a \neq 0\}$

17. $\{\text{reals}, x \neq 7\}$ **19.** $\{\text{reals}, x \neq -5\}$ **21.** $\{\text{all reals}\}$ **23.** $\{\text{reals}, x \neq -2 \text{ or } 2\}$ **25.** $\dfrac{5}{3a^2}$

27. $\dfrac{2s}{3r}$ **29.** $\dfrac{6b}{5a}$ **31.** $\dfrac{8qr^5}{7p^3}$ **33.** $\dfrac{7}{8}$ **35.** $\dfrac{b}{2}$ **37.** $\dfrac{k - 3}{k + 5}$ **39.** $\dfrac{x + 1}{x - 3}$ **41.** $\dfrac{a + 3}{a}$

43. $\dfrac{1}{x - 5}$ **45.** $\dfrac{x - 1}{8}$ **47.** $\dfrac{x + y}{x - y}$ **49.** $\dfrac{-2}{3}$ **51.** -1 **53.** $-(4 + r)$ **55.** $\dfrac{-(3 + x)}{x}$

57. $\dfrac{x - 6}{x - 5}$ **59.** $\dfrac{m + 4}{m - 5}$ **61.** $\dfrac{t + 7}{t + 5}$ **63.** $\dfrac{x - 5}{x + 7}$ **65.** $16(t + 3)$

PROBLEM SET 5.2

1. $\dfrac{8}{33}$ **3.** $\dfrac{1}{8}$ **5.** $\dfrac{4}{11}$ **7.** $\dfrac{-1}{12}$ **9.** $\dfrac{8}{9}$ **11.** $\dfrac{40}{63}$ **13.** $\dfrac{2y}{3x^4}$ **15.** $\dfrac{2p}{3mn^2q}$

17. $\dfrac{x^3}{8a^3b^2cy^2z^3}$ **19.** $a(a - 5)$ **21.** $\dfrac{k - 2}{k - 6}$ **23.** $\dfrac{2}{3}$ **25.** $\dfrac{x^4(x - 5)}{(x - 6)(x + 6)}$ **27.** $\dfrac{m - 9}{m - 5}$

29. $\dfrac{u + 1}{u + 2}$

PROBLEM SET 5.3

1. $\dfrac{7}{3}$ **3.** $\dfrac{-3}{5}$ **5.** $\dfrac{1}{12}$ **7.** $\dfrac{v}{u}$ **9.** $\dfrac{z}{x^2 + y^2}$ **11.** $\dfrac{k^2 - 9}{k^2 - 25}$ **13.** $\dfrac{21}{10}$ **15.** $\dfrac{1}{6}$ **17.** $\dfrac{-5}{6}$

19. $\dfrac{-9}{5}$ **21.** $\dfrac{3}{14}$ **23.** $\dfrac{1}{ba^2}$ **25.** $\dfrac{x^4 r^2}{6y^2 s}$ **27.** $\dfrac{2c^4 x^2 z}{3ab}$ **29.** $\dfrac{2}{x^5}$ **31.** $\dfrac{5}{2a}$ **33.** $\dfrac{m}{m + 3}$

35. $\dfrac{p + 4}{p - 3}$ **37.** $\dfrac{t^2 + 13t + 36}{t^2 - 4t - 60}$ **39.** $\dfrac{r - 2}{4r - 3}$

PROBLEM SET 5.4

1. $\dfrac{5}{9}$ **3.** $\dfrac{2}{3}$ **5.** $\dfrac{13}{10}$ **7.** $\dfrac{8}{13}$ **9.** $\dfrac{-1}{5}$ **11.** $\dfrac{9}{16}$ **13.** $\dfrac{5}{x}$ **15.** $\dfrac{3x - 1}{a + b}$ **17.** $\dfrac{7u - 4}{u(u + 7)}$

19. $\dfrac{x^2 + 4x - 6}{x(x + 8)}$ **21.** $\dfrac{5}{t}$ **23.** $\dfrac{-x + 3}{x + y}$ **25.** $\dfrac{3k - 5}{k(k + 2)}$ **27.** $\dfrac{-13}{t(t - 5)}$ **29.** 1

PROBLEM SET 5.5

1. $\dfrac{5}{6}$ **3.** $\dfrac{3}{10}$ **5.** $\dfrac{-13}{18}$ **7.** $\dfrac{43}{36}$ **9.** $\dfrac{67}{168}$ **11.** $\dfrac{61}{126}$ **13.** $\dfrac{a + b}{ab}$ **15.** $\dfrac{6 - m^2}{2m}$

17. $\dfrac{6 - 5b}{3ab}$ **19.** $\dfrac{8x + 6}{3x^2}$ **21.** $\dfrac{2ax - by}{4x^2 y^2}$ **23.** $\dfrac{4uw^2 + 5v^2}{6u^2 v^2 w^2}$ **25.** $\dfrac{29}{6x - 18}$ **27.** 0

29. $\dfrac{11x - 34}{(x + 2)(x - 5)}$ **31.** $\dfrac{5k + 11}{(2k + 3)(k + 2)}$ **33.** $\dfrac{5x + 21}{3(x - 3)(x + 3)}$ **35.** $\dfrac{8a + 4}{a(a + 2)(a - 2)}$

37. $\dfrac{8a - 38}{(a - 4)(a - 5)(a - 6)}$ **39.** $\dfrac{-2}{(x - 1)(x - 2)}$ **41.** $\dfrac{m - 50}{(m + 6)(m + 6)(m - 1)}$ **43.** $\dfrac{-9}{(x - 7)(x + 4)}$

45. $\dfrac{3m^2 + 9m + 7}{(m + 7)(m + 1)(2m + 1)}$ **47.** $\dfrac{-10{,}000h}{p(p + h)}$ **49.** $\dfrac{1}{f} = \dfrac{(n - 1)(R_2 - R_1)}{R_1 R_2}; f = \dfrac{R_1 R_2}{(n - 1)(R_2 - R_1)}$

PROBLEM SET 5.6

1. $\dfrac{14}{15}$ **3.** $\dfrac{6}{7}$ **5.** $\dfrac{9x}{8}$ **7.** $\dfrac{ab}{c}$ **9.** 6   **11.** $\dfrac{16}{27}$ **13.** $\dfrac{-5}{13}$ **15.** $\dfrac{69}{35}$ **17.** $\dfrac{b - a}{b + a}$

19. $\dfrac{2xy - y}{3xy + x}$ **21.** $\dfrac{6a^2 + a}{7a^2 - 1}$ **23.** $\dfrac{1 - 2b}{2ab + 3a}$ **25.** $\dfrac{-P(\Delta Q)}{Q(\Delta P)}$

PROBLEM SET 5.7

1. 3   **3.** 2   **5.** 6   **7.** 8   **9.** 10   **11.** 4   **13.** -2 **15.** 3   **17.** 15   **19.** 7
21. 2   **23.** -3 **25.** No solution   **27.** 7   **29.** 6   **31.** 60   **33.** 30

PROBLEM SET 5.8

1. 19   **3.** 10   **5.** 6   **7.** $\dfrac{15}{4}$ **9.** 12   **11.** 10   **13.** 30(Jerry), 40(Eddie)

15. 20(Jennifer), 15(Melissa)   **17.** 500(car), 2500(train)

CHAPTER 5 REVIEW EXERCISES

1. $\dfrac{3}{7}$ **2.** $\dfrac{6}{5}$ **3.** $\dfrac{3s^2}{2r^2}$ **4.** $\dfrac{5}{t}$ **5.** $\dfrac{x-5}{2}$ **6.** $\dfrac{-1}{k+3}$ **7.** $\dfrac{a-5}{a-4}$ **8.** {all reals}

9. {reals, $a \neq 6$}   **10.** $\dfrac{7}{3a^2}$ **11.** $\dfrac{3x+7}{2x-5}$ **12.** $\dfrac{3}{8}$ **13.** $\dfrac{-5}{6}$ **14.** $\dfrac{5d}{6ab^2c^3}$

15. $\dfrac{(t+5)(t+2)}{t^3}$ **16.** $\dfrac{x+5}{x+7}$ **17.** $\dfrac{4}{3}$ **18.** $\dfrac{2bd^4}{3ac}$ **19.** $\dfrac{a+2}{a^4}$ **20.** $\dfrac{t-3}{t-6}$ **21.** $\dfrac{16}{17}$

22. $\dfrac{1}{11}$ **23.** $\dfrac{8r-3}{(r-3)(r+4)}$ **24.** $\dfrac{k+2}{a+b}$ **25.** $\dfrac{1}{6}$ **26.** $\dfrac{34}{175}$ **27.** $\dfrac{14+4a}{10a^2}$

28. $\dfrac{-2a-37}{(a+5)(a-4)}$ **29.** $\dfrac{2x^2+6x-12}{(x-4)(x+1)(x-3)}$ **30.** $\dfrac{3}{2a}$ **31.** $\dfrac{-2}{3}$ **32.** $\dfrac{a-1}{5a^2+3a}$ **33.** 2

34. 5   **35.** 4   **36.** 7   **37.** $\dfrac{21}{10}$ **38.** 45(Eric), 55(Matthew)

PROBLEM SET 6.1

1. $(1, 9); (-4, 14)$ **3.** $(3, 4)$ **5.** $(0, -4); (2, 1)$ **7.** $(3, -1); (17, 1); (10, 0)$ **9.** $(4, 6); (13, -3)$
11. $(0, 6); (6, -6)$ **13.** $(1, 2); (0, -3)$ **15.** $(6, 0); (3, 4)$ **17.** $(5, -6); (12, 15)$
19. $(-1, -8); (2, 16)$ **21.** $(5, 7); (5, -3)$ **23.** $(2, 10); (-8, 10)$

25.

x	0	4	-4	8
y	-3	0	-6	3

27.

x	0	12	30	24
y	-2	0	3	2

29.

x	4	4	4	4
y	0	5	-7	12

31. $(0, 33); (5, 22); (15, 0)$ **33.** $(2, 4000); (5, 1000); (3.5, 2500)$

PROBLEM SET 6.2

1–15.

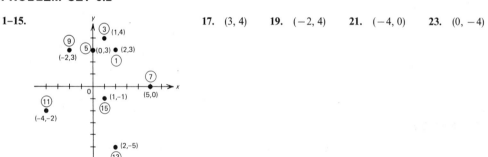

17. $(3, 4)$ **19.** $(-2, 4)$ **21.** $(-4, 0)$ **23.** $(0, -4)$

25. $(30, 50)$ **27.** $(0, 200)$ **29.** $(-40, 100)$ **31.** $(-20, -150)$ **33.** $(40, -200)$

PROBLEM SET 6.3

1. $(0, 4); (2, 2); (4, 0)$ **3.** $(0, 10); (3, 4); (5, 0)$ **5.** $(0, -2); (18, 1); (12, 0)$

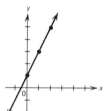      

7. $(0, 1); (1, 3); (2, 5)$ **9.** $(3, 4); (3, -1); (3, 0)$ **11.**

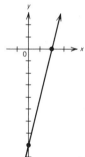   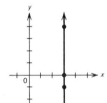   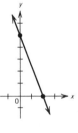

13.   **15.**   **17.**

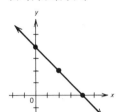

19.   **21.**   **23.**

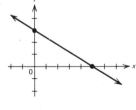   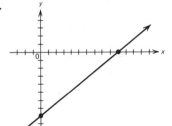   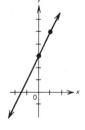

25. 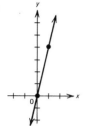   **27.** 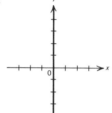   **29.**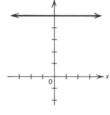

31. 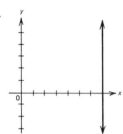   **33.** 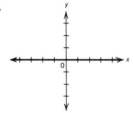   **35.**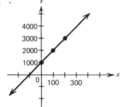

37. 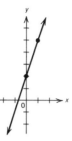   **39.**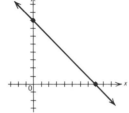

PROBLEM SET 6.4

1. (9, 3) is a solution   **3.** (1, −3) is a solution   **5.** None is a solution

7. 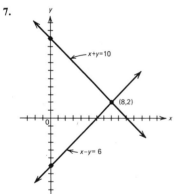   **9.**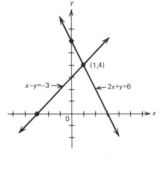

11.

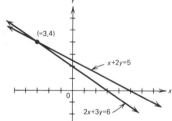

$(=3,4)$

$x+2y=5$

$2x+3y=6$

13.

Parallel
(no solution)

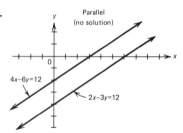

$4x-6y=12$

$2x-3y=12$

15.

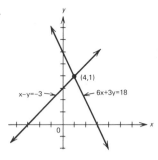

$(4,1)$

$x-y=-3$

$6x+3y=18$

17.

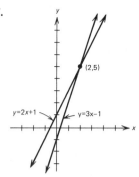

$(2,5)$

$y=2x+1$

$y=3x-1$

19.

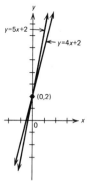

$y=5x+2$

$y=4x+2$

$(0,2)$

21.

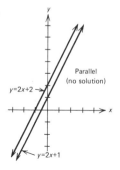

Parallel
(no solution)

$y=2x+2$

$y=2x+1$

23.

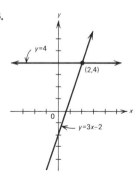

$y=4$

$(2,4)$

$y=3x-2$

25.

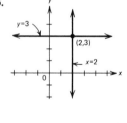

$y=3$

$(2,3)$

$x=2$

27.

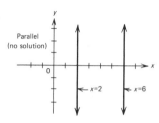

Parallel
(no solution)

$x=2$

$x=6$

29.

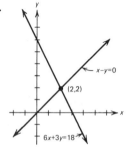

$x-y=0$

$(2,2)$

$6x+3y=18$

PROBLEM SET 6.5

1. (6, 4)   **3.** (2, 5)   **5.** (−3, −9)   **7.** (2, 3)   **9.** (13, 4)   **11.** (7, 1)   **13.** (1, −4)
15. (6, 1)   **17.** (1, −5)   **19.** (0, 3)   **21.** (1, 3)   **23.** (1, −2)   **25.** (−1, −2)
27. Either equation   **29.** (0, −2)   **31.** (7, 1)   **33.** (5, 2)   **35.** (−2, 4)   **37.** (100, 50)

PROBLEM SET 6.6

1. (1, 3)   **3.** (−1, 6)   **5.** (3, 5)   **7.** No solution   **9.** (−2, −1)   **11.** (−2, −7)
13. (−4, −5)   **15.** (2, 0)   **17.** (−1, −5)   **19.** (12, 2)   **21.** (52, 13)   **23.** (1, 3)   **25.** (4, 2)
27. No solution   **29.** (7, 3)   **31.** (20,000, 2800)   **33.** (10, 9000)

PROBLEM SET 6.7

1. 29, 18   **3.** 8, 48   **5.** 6, 11   **7.** 3500 at 5%; 1500 at 8%   **9.** 1500 at 7%; 3500 at 8%
11. 20 of 40%; 60 of 20%   **13.** 30 at 90%; 10 at 70%   **15.** 600 at \$4; 400 at \$5;
17. 30 at \$3 per pound; 10 at \$4 per pound

CHAPTER 6 REVIEW EXERCISES

1. (1, 1); (6, −1); (−4, 3)   **2.** (3, 2); (0, 4)   **3.** (10, 0); (15, 4)   **4.** (10, 3); (10, −2)

5.

x	0	7	−7	14
y	−2	0	−4	2

6–7.

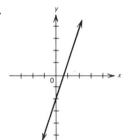

8. (0, 2)   **9.** (3, −1)

10.

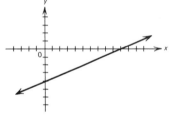

11.

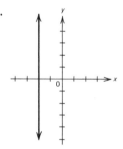

12.

13.

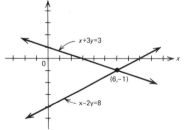

14.

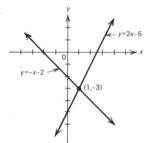

15.

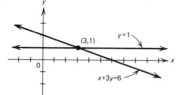

16. $(2, 5)$ **17.** $(-2, 3)$ **18.** $(2, 7)$ **19.** $(3, -5)$

20. $(4, -3)$ **21.** No solution   **22.** 39, 23   **23.** 1300 at 7%; 700 at 8%   **24.** 180 at \$5; 120 at \$3

PROBLEM SET 7.1

1. $9, -9$ **3.** $4, -4$ **5.** No real number   **7.** $8, -8$ **9.** $12, -12$ **11.** 5   **13.** 0
15. -8 **17.** No real number   **19.** -2 **21.** 30   **23.** $7.348, -7.348$ **25.** $1.414, -1.414$
27. 4.243   **29.** -7.810 **31.** No real number   **33.** $10; -10$; rational
35. $3.464; -3.464$; irrational   **37.** No real number   **39.** $16; -4$; rational   **41.** $9; 3$; rational
43. $6; -\sqrt{6}$; irrational   **45.** 2   **47.** 4   **49.** 3   **51.** 2   **53.** 10
55. (a) 30,000 (b) 100,000 (c) 40,000   **57.** (a) 100 (b) 77.5 (c) 71.7

PROBLEM SET 7.2

1. $\sqrt{21}$ **3.** 9   **5.** 10   **7.** 5   **9.** 10   **11.** $5\sqrt{2}$ **13.** $3\sqrt{2}$ **15.** $10\sqrt{10}$ **17.** $7\sqrt{2}$

19. $4\sqrt{2}$ **21.** $10\sqrt{2}$ **23.** $5\sqrt{6}$ **25.** 30   **27.** $20\sqrt{6}$

29. $10\sqrt{6}$ **31.** $20\sqrt{14}$ **33.** $\dfrac{9}{10}$ **35.** $\dfrac{5}{6}$ **37.** $\dfrac{7}{3}$ **39.** $\dfrac{11}{12}$ **41.** $\dfrac{\sqrt{7}}{5}$ **43.** $\dfrac{\sqrt{3}}{7}$

45. 0.7   **47.** 0.09   **49.** $\dfrac{3}{2}$ **51.** $\dfrac{5}{2}$ **53.** $\dfrac{1}{2}$ **55.** $\dfrac{\sqrt{15}}{2}$ **57.** $\dfrac{5\sqrt{3}}{7}$ **59.** $4x^5$

61. $5p^6\sqrt{2}$ **63.** $2t^2\sqrt{6t}$ **65.** $\dfrac{x}{3}$ **67.** $\dfrac{u^2}{4}\sqrt{u}$ **69.** 50

PROBLEM SET 7.3

1. $9\sqrt{2}$ **3.** $4\sqrt{3}$ **5.** $5\sqrt{10}$ **7.** $-7\sqrt{11}$ **9.** $9 + 5\sqrt{2}$ **11.** $5 - 2\sqrt{7}$ **13.** $7\sqrt{5}$
15. $4\sqrt{7} - 9\sqrt{2}$ **17.** $5\sqrt{5}$ **19.** $6\sqrt{2}$ **21.** $-2\sqrt{5}$ **23.** $7\sqrt{5}$ **25.** $13\sqrt{2x}$
27. $(1 + 2a)\sqrt{5a}$ **29.** $(1 - 2x + 5x^2)\sqrt{x}$ **31.** $(5 - 8u + 10u^2)\sqrt{2u}$ **33.** $\sqrt{15} + 3$
35. $\sqrt{30} - 2\sqrt{3}$ **37.** 15   **39.** $20 - 5\sqrt{3}$ **41.** $16 + 9\sqrt{3}$ **43.** -12 **45.** $-1 - \sqrt{6}$
47. $7 + \sqrt{15}$ **49.** $3 + 2\sqrt{2}$ **51.** $31 - 10\sqrt{6}$ **53.** $9 + 2\sqrt{14}$ **55.** $13 - 2\sqrt{22}$ **57.** -2
59. -2 **61.** -1 **63.** 1   **65.** 0

PROBLEM SET 7.4

1. $\dfrac{5\sqrt{6}}{6}$ **3.** $\dfrac{8\sqrt{3}}{3}$ **5.** $8\sqrt{2}$ **7.** $\dfrac{5\sqrt{3}}{3}$ **9.** $\dfrac{\sqrt{22}}{11}$ **11.** $\dfrac{\sqrt{2}}{2}$ **13.** $\dfrac{\sqrt{22}}{11}$ **15.** $\dfrac{\sqrt{91}}{13}$

17. $\dfrac{2\sqrt{51}}{17}$ **19.** $\dfrac{\sqrt{ab}}{b}$ **21.** $\dfrac{\sqrt{3xy}}{3y}$ **23.** $\dfrac{\sqrt{10rs}}{5s^2}$ **25.** $\dfrac{a\sqrt{70ab}}{7b^3}$ **27.** $\dfrac{12 - 4\sqrt{2}}{7}$ **29.** $2 + \sqrt{3}$

31. $\dfrac{2\sqrt{2} + 14}{-47}$ **33.** $\dfrac{\sqrt{22} - 3\sqrt{2}}{2}$ **35.** $\sqrt{3} + \sqrt{2}$ **37.** $\dfrac{4\sqrt{7} - 4\sqrt{2}}{5}$ **39.** $\sqrt{5} + \sqrt{2}$

41. $\dfrac{\sqrt{6} - \sqrt{2}}{2}$ **43.** $\dfrac{10u - 10\sqrt{u}}{u - 1}$ **45.** $P = \dfrac{600\sqrt{h}}{h}$

PROBLEM SET 7.5

1. 24   **3.** -96 **5.** 4   **7.** No solution   **9.** -5 **11.** 2   **13.** 2   **15.** No solution
17. 2   **19.** 1, -1 **21.** 2, -2 **23.** No solution   **25.** 10, -10 **27.** 0, 10   **29.** 11, 9
31. 4, -7 **33.** 5, -2 **35.** $-5 + \sqrt{6}, -5 - \sqrt{6}$ **37.** $4 + \sqrt{3}, 4 - \sqrt{3}$

39. $\dfrac{-1 + \sqrt{11}}{2}, \dfrac{-1 - \sqrt{11}}{2}$ **41.** No solution   **43.** 268.7 feet

PROBLEM SET 7.6

1. $-5, -2$ **3.** $6, -3$ **5.** $-1, -4$ **7.** $-3, 2$ **9.** 1, 10   **11.** $\dfrac{-1}{2}, \dfrac{1}{3}$ **13.** $\dfrac{-5}{2}, \dfrac{-1}{4}$

15. $\dfrac{1}{2}, -3$ **17.** $\dfrac{-9 \pm \sqrt{97}}{2}$ **19.** $\dfrac{3 \pm \sqrt{29}}{10}$ **21.** $\dfrac{1 \pm \sqrt{61}}{6}$ **23.** $\dfrac{-1 \pm \sqrt{41}}{10}$ **25.** $4 \pm \sqrt{14}$

27. $\dfrac{-2 \pm \sqrt{10}}{2}$ **29.** $4, -1$ **31.** $\sqrt{5}, -\sqrt{5}$ **33.** $\dfrac{3 \pm \sqrt{33}}{2}$ **35.** $\dfrac{5 \pm 3\sqrt{5}}{2}$ **37.** $\dfrac{-1 \pm \sqrt{11}}{2}$

39. $0.424, -9.424$ **41.** $-1.135, 1.468$ **43.** $-2.581, 0.581$ **45.** $-1.372, 4.372$

47. $\dfrac{1 + 2\sqrt{7}}{3} \approx \$2.10$ **49.** $\dfrac{1 + \sqrt{5}}{2} \approx 1.618$

CHAPTER 7 REVIEW EXERCISES

1. $-9, 9$ **2.** No real number   **3.** 6   **4.** -7 **5.** 3   **6.** 2   **7.** Rational

8. Irrational   **9.** 6   **10.** $5\sqrt{5}$ **11.** 24   **12.** $\dfrac{\sqrt{11}}{4}$ **13.** 0.4   **14.** $\dfrac{\sqrt{5}}{2}$ **15.** $\dfrac{x\sqrt{x}}{5}$

16. $-10\sqrt{5}$ **17.** $-1 + 10\sqrt{2}$ **18.** $7\sqrt{3}$ **19.** $(1 - 3x)\sqrt{3x}$ **20.** $\sqrt{35} - \sqrt{10}$

21. $-1 + 2\sqrt{10}$ **22.** $7 - 4\sqrt{3}$ **23.** 23   **24.** $\dfrac{\sqrt{35}}{7}$ **25.** $\dfrac{\sqrt{77}}{11}$ **26.** $\dfrac{\sqrt{6xy}}{3y^2}$ **27.** $\dfrac{15 + 3\sqrt{2}}{23}$

28. $\dfrac{\sqrt{14} - \sqrt{6}}{4}$ **29.** 27   **30.** 1   **31.** $9, -9$ **32.** $\dfrac{3}{4}, \dfrac{-1}{4}$ **33.** $\dfrac{3 \pm \sqrt{7}}{2}$ **34.** $-2, \dfrac{-3}{2}$

35. $\dfrac{5 \pm \sqrt{73}}{8}$ **36.** $3 \pm \sqrt{5}$ **37.** $\dfrac{1 \pm \sqrt{37}}{2}$ **38.** $5.236, 0.764$

PROBLEM SET 8.1

1. $\dfrac{7}{6}$ **3.** $\dfrac{-1}{3}$ **5.** $\dfrac{1}{2}$ **7.** $\dfrac{7}{5}$ **9.** $\dfrac{-1}{2}$ **11.** Undefined   **13.** 0

15. 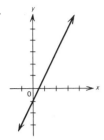   **17.** 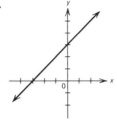   **19.**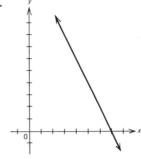

21. 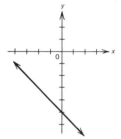   **23.** **(a)** 200 **(b)** 500 **(c)** -200

PROBLEM SET 8.2

1. $y = 2x + 3$ **3.** $y = 7x + 19$ **5.** $y = -3x + 17$ **7.** $y = -10x - 58$ **9.** $y = -1$

11. $y = 2x + 3$ **13.** $y = 4x - 5$ **15.** $y = 4x + 22$ **17.** $y = 8$ **19.** $m = 7; b = 3$

21. $m = -3; b = 4$ **23.** $m = 6; b = -8$ **25.** $m = 3; b = 0$ **27.** $m = 0; b = 8$

29. $m = -2; b = 5$ **31.** $m = -2; b = \dfrac{3}{4}$ **33.** $m = \dfrac{3}{2}; b = \dfrac{-5}{2}$ **35.** $m = \dfrac{-4}{5}; b = 0$

37. 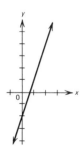   **39.** 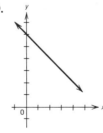   **41.**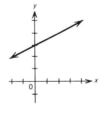

43. $y = \dfrac{-1}{200} x + 45$ **45.** **(a)** $y = 0.08x + 1200$ **(b)** $0.08\dfrac{\$}{\text{mile}}$ **(c)** \$1200

PROBLEM SET 8.3

1.
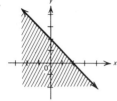

3.
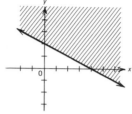

5.
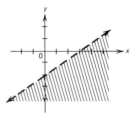

7.
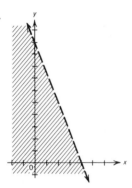

9.
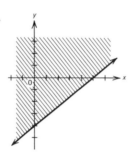

11.
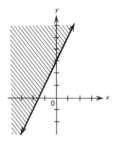

13.
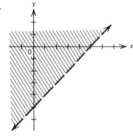

15.
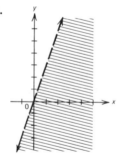

17.
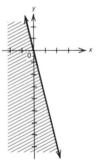

19.
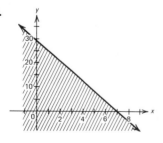

21.
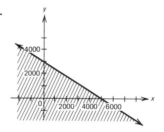

PROBLEM SET 8.4

1.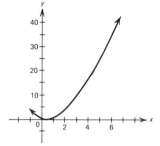

3.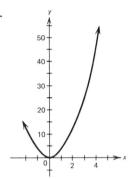

5.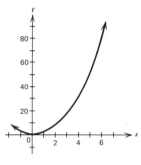

7.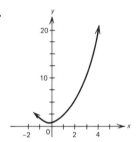

9.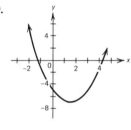

11.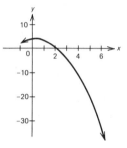

13.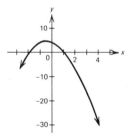

15. (a)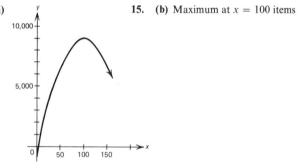

15. (b) Maximum at $x = 100$ items

17. (a)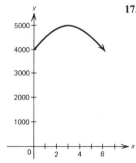

17. (b) Maximum at $x = 3$ hours

CHAPTER 8 REVIEW EXERCISES

1. 4   **2.** −1   **3.** $\dfrac{3}{2}$ **4.** $\dfrac{-2}{7}$

5. 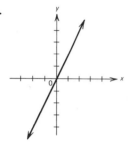   **6.** 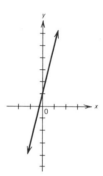   **7.**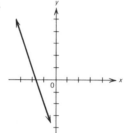

8. 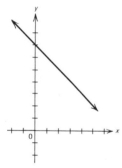   **9.** $y = 3x - 14$ **10.** $y = -5x - 13$ **11.** $y = 4x - 1$

12. $y = -3x$ **13.** 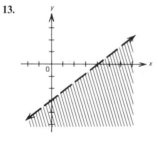   **14.**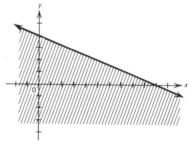

15. 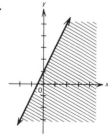   **16.** 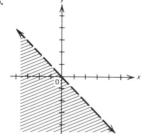   **17.**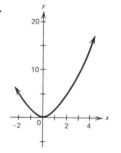

18.

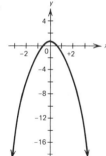

19.

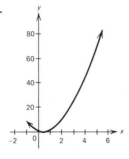

INDEX
OF MATHEMATICAL TERMS

INDEX
OF APPLICATIONS